A Visual Analogy Guide to Human Anatomy

Second Edition

Paul A. Krieger

Grand Rapids Community College

Morton Publishing Company
925 W. Kenyon Ave., Unit 12
Englewood, CO 80110
800-348-3777
http://www.morton-pub.com

BOOK TEAM

Publisher	Douglas N. Morton
Biology Editor	David Ferguson
Cover & Design	Bob Schram, Bookends, Inc.
Composition	Ash Street Typecrafters, Inc.
Illustration	Paul Krieger

Printed in the United States of America

10 9 8 7 6 5 4 3 2 1

ISBN-10: 0-89582-800-6
ISBN-13: 978-089582-800-2

For my wife, Lily

For your love, constant support, and endless encouragement
Without your help this book would
never have been written

And for my son, Ethan

May you never lose that sparkle in your eye and
May you always follow your passions in life

Acknowledgments

So many people contributed to this book that I would like to acknowledge them. My editor, David Ferguson, for his support and thoughtfulness in giving me the time and freedom to complete the book I originally envisioned. To Dan Matusiak, for his diligence in faithfully creating the glossary. I would also like to offer a special thank you to reviewers Cynthia Herbrandson, Chris Sullivan, and David Canoy for their valuable feedback during the writing process. My wife, Lily Krieger, kindly accepted all my long working hours away from the family. Without her help and support this book would not have been written. Thanks to Mike Timmons, for helping me crystallize my initial ideas about the visual analogy guides and turn them into a tangible reality. To my friend, Kevin Patton, for graciously responding to all my authoring questions along the way. Finally, for my students, friends in the Human Anatomy & Physiology Society (HAPS), and everyone else who offered suggestions, support, and encouragement. Thanks to all of you. I am truly grateful.

Contents

Muscular System 123

Nervous System 141

Endocrine System 161

Special Senses 165

Blood 177

Cardiovascular System 183

How to Use this Book

Purpose

This book was written primarily for students of human anatomy; however, it will be useful for teachers or anyone else with an interest in this topic. It was designed to be used in conjunction with any of the major anatomy or anatomy and physiology textbooks. What makes it unique, creative, and fun is the **visual analogy** learning system. This will be explained later. The modular format allows you to focus on one key concept at a time. Each module has a text page on the left with corresponding illustrations on the facing page. Most illustrations are unlabeled so that you can quiz yourself on the structures. A handy key to the illustration is provided on the text page. While this book covers most all major organ systems, the topics are weighted more toward areas that typically give students difficulty. It uses a variety of learning activities such as labeling, coloring, and mnemonics to help instruct. In addition, it offers special study tips for mastering difficult topics.

What Are *Visual Analogies*?

A visual analogy is a helpful way to learn new material based on what you already know from everyday life. It compares an anatomical structure to something familiar such as an animal or a common object. For example, the vertebral column has three different types of vertebrae. One type looks like a giraffe. Comparing the vertebra to a giraffe allows you to mentally correlate the *unknown* (*vertebra*) with the *known* (*giraffe*). Doing this accomplishes several things.

1. It reduces your anxiety about learning the material and helps you focus on the task at hand.
2. It forces you to observe anatomical structures more carefully. After all, being a good observer is the first step to becoming a good anatomist—or any type of scientist!
3. It makes the learning more fun, relevant, and meaningful so you can better retain the information.

Whenever a visual analogy is used in this book, a small picture of it appears in the upper righthand corner of the illustration page for easy reference. This allows you to quickly reference a page visually, simply by flipping through the pages.

Icons Used

The following icons are used throughout this book:

 Microscope icon—Indicates any illustration that is microscopic.

 Crayon icon—Indicates illustrations that were especially made for coloring. Even though you may color any of the illustrations to enhance your learning, it may be more beneficial to do so with those that have this icon. In some cases, written instructions may appear next to this icon with directions about exactly *what* to color or *how* to color it.

 Three-dimensional icon—Indicates a three-dimensional view of an anatomical structure.

 Two-dimensional icon—Indicates a two-dimensional view of an anatomical structure.

Abbreviations and Symbols

The following abbreviations/symbols are commonly used throughout this book:

l. = left   v. = vein

r. = right   m. = muscle

a. = artery   n. = nerve

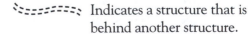 Indicates a structure that is behind another structure.

It is my sincere hope that this book will be a fun, effective study tool for anyone interested in learning human anatomy. Many of the visual analogies used in the book have been tested repeatedly on students in the anatomy lab and/or lecture to ensure that they are useful.

Enjoy learning with visual analogies!

VISUAL ANALOGY INDEX

TOPIC	ANALOGY	ICON(S)	PAGE NO.
1. Language of Anatomy—Body Cavities and Membranes	A serous membrane is like a **fist in a balloon**		15
2. Cells—Plasma Membrane	A plasma membrane is like a **waterbed**		23
3. Tissues—Simple Squamous Epithelium	Each cell looks like a **fried egg**		29
4. Tissues—Simple Cuboidal Epithelium	Each cell looks like an **ice cube**		31
5. Tissues—Simple Columnar Epithelium	Each cell looks like a **column**		33
6. Tissues—Variations in Connective Tissues	Collagen is like a **steel cable**; elastin is like a **rubberband**		41, 49
7. Tissues—Reticular Connective Tissues	Reticular fibers are like **cobwebs**		47
8. Tissues—Cartilage	Cartilage is like a block of **Swiss cheese**		53, 55, 57
9. Tissues—Bone	Osteon looks like a **tree stump**		59, 77
10. Tissues—Skeletal Muscle	Each skeletal muscle cell looks like a **birch tree log**		61
11. Tissues—Cardiac Muscle	The intercalated disc is like **two pieces of a jigsaw puzzle**		63
12. Tissues—Smooth Muscle	Smooth muscle stack in a sheet like **bricks in a wall**		65
13. Tissues—Nervous Tissue	Multipolar neuron looks like an **octopus**		67

VISUAL ANALOGY INDEX

TOPIC	ANALOGY	ICON(S)	PAGE NO.
14. Skeletal System—Skull	The coronal suture is like a **tiara (crown)** The sutures on the posterior aspect of the skull look like a **modified peace sign**		80
15. Skeletal System—Skull	The sella turcica of the sphenoid bone looks like a **horse's saddle**		81
16. Skeletal System—Temporal Bone	The temporal bone resembles a **rooster's head**		85
17. Skeletal System—Ethmoid Bone	Superior view looks like a **door hinge**; crista galli looks like a **shark fin**		87
18. Skeletal System—Sphenoid Bone	Sphenoid resembles a **bat**		89
19. Skeletal System—Sphenoidal Foramina	Remember the sphenoidal foramina with **Ros the cowboy**		91
20. Skeletal System—Palatine Bones	Palatine bones are like **2 letter "L"s**—one the mirror image of the other		93
21. Skeletal System—Numbers of Vertebrae	Total number of each type of vertebrae correspond to **meal times**		95
22. Skeletal System—Atlas and Axis	Atlas the **turtle head**; Axis the **football player**		97
23. Skeletal System—Lumbar Versus Thoracic	Thoracic and lumber vertebrae are like a goose with **wings in different positions**		99
24. Skeletal System—Lumbar Versus Thoracic	**"Thoracic giraffe"**; **"Lumbering moose"**		101

VISUAL ANALOGY INDEX

TOPIC	ANALOGY	ICON(S)	PAGE NO.
25. Skeletal System—Humerus	Distal end of the humerus looks like the **hand of a hitchhiker**		107
26. Skeletal System—Radius and Ulna	Head of radius—**hockey puck**; Ulna—**crescent wrench**		109
27. Skeletal System—Pelvis	Pubis bones—**mask**; Coccyx—**rattlesnake tail**		113 95
28. Muscular System—Actin Filament	Each actin filament is like a **double-stranded chain of pearls**		125
29. Muscular System—Muscles of Neck, Shoulder, Thorax, and Abdomen	Pectoralis major—**fan**; Serratus anterior—**serrated knife**; Abdominal muscles—**sandwich**		131
30. Muscular System—Muscles of Thigh	Sartorius is like a **sash**		135
31. Muscular System—Muscles that Move Ankle, Foot, and Toes	Soleus is like a **sole flatfish**		137
32. Nervous System—Multipolar Neuron	Axon of a neuron is like an **electrical cord**		145
33. Nervous System—Peripheral Nerve	A peripheral nerve is like **tubes within tubes**		147
34. Nervous System—Spinal Cord	The gray matter in the spinal cord is shaped like a **butterfly**		149
35. Nervous System—Brain Ventricles	The ventricular system is like the **neck, head, and horns of a ram**		157
36. Special Senses—Eye—Internal	Macula lutea is like a **target**; Fovea centralis—**bullseye**		169

VISUAL ANALOGY INDEX

	TOPIC	ANALOGY	ICON(S)	PAGE NO.
37.	Special Senses—Ear—General Structure	Malleus—**hammer**; Incus—**anvil** Stapes—**stirrup** Cochlea—**snail shell**		171
38.	Special Senses—Tongue	Fungiform papilla—**mushroom cap** Filiform papilla—**flame**		173
39.	Cardiovascular System—Heart	A-V valves—**parachute**; Valve flap—**kangaroo pouch** Semilunar valves—**modified peace sign**		187
40.	Lymphatic System—Creaton of Lymph	Lymph nodes filter debris like an **oil filter** filters oil in a car engine		217
41.	Respiratory System—Overview	Larynx—**head of snapping turtle**; Alveoli—**bubblewrap**		227
42.	Digestive System—Small Intestine	Plicae circularis—**folded carpet sample**; Villus—**single carpet fiber**		241
43.	Digestive System—Pancreas	Pancreas—**tadpole**		243
44.	Digestive System—Liver	Hepatic lobule is like a **ferris wheel**		245
45.	Digestive System—Appendix	Appendix is like a **worm**		249
46.	Urinary System—Kidney	The calyces are like a **plumbing system**		255
47.	Urinary System—Nephron Structure	The collecting tubule is like a **public sewer drain**		257
48.	Male Reproductive System—Penis	The penis in cross-section looks like a **monkey face**		263
49.	Female Reproductive System—Ovary	Ovulation is like a **water balloon bursting**		271

Language
of
Anatomy

Description

An essential skill in anatomy is being able to visualize a sliced section of a tissue, organ, or region of the human body. This requires you to mentally jump from the three-dimensional to the two-dimensional. The prerequisite to developing this skill is being able to visualize the different ways that an object can be sliced. The three basic planes that can pass through an object to section it are:

- **Sagittal** (*median*) **plane**—This plane slices an object down the middle, making a left half and a right half.

- **Frontal plane**—This is also called the coronal plane. The term coronal means "crown." This plane splits an object into a front half and a back half.

- **Transverse plane**—This plane divides the body into an upper half and a lower half.

Key to Illustration

1. Frontal plane   2. Sagittal (*median*) plane   3. Transverse plane

**Planes of the Body
(through a 2-year old boy)**

1. _____

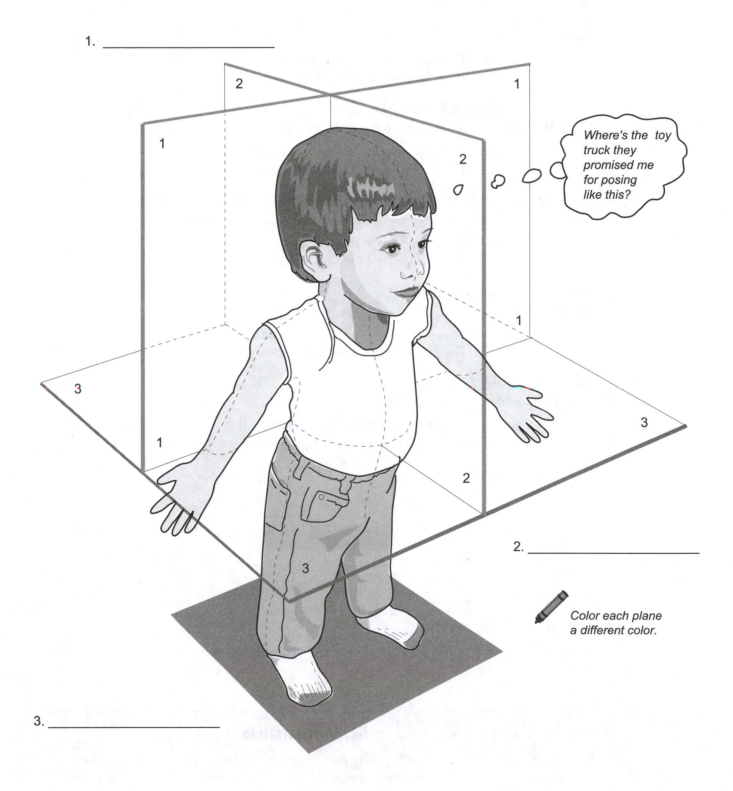

Where's the toy truck they promised me for posing like this?

2. _____

Color each plane a different color.

3. _____

Description

Directional terms are part of the working language of anatomy. They are commonly used to describe the position of an anatomical structure or the position of one body part in relation to another. Some of the most common terms for body orientation/direction are given below, along with their meanings and examples of how to use them correctly in a sentence.

Superior: above
Inferior: below

> *ex:* The brain is *superior* to the lungs.
> The lungs are *inferior* to the brain.

Medial: toward the midline of the body
Lateral: away from the midline of the body

> *ex:* The nose is *medial* to the external ear.
> The external ear is *lateral* to the nose.

Proximal: nearer the trunk of the body or point of attachment
Distal: farther from the trunk of the body or point of attachment

> *ex:* The shoulder is *proximal* to the wrist.
> The wrist is *distal* to the shoulder.

Superficial: toward the body surface
Deep: away from the body surface

> *ex:* The skin is *superficial* to the muscle.
> The muscles are *deep* to the skin.

Anterior/ventral: toward the front
Posterior/dorsal: toward the back

> *ex:* The heart is *anterior* to the spinal cord.
> The spinal cord is *posterior* to the heart.

Key to Illustration

1. Superior	3. Anterior (ventral)	5. Proximal	7. Medial
2. Inferior	4. Posterior (dorsal)	6. Distal	8. Lateral

Directional Terms

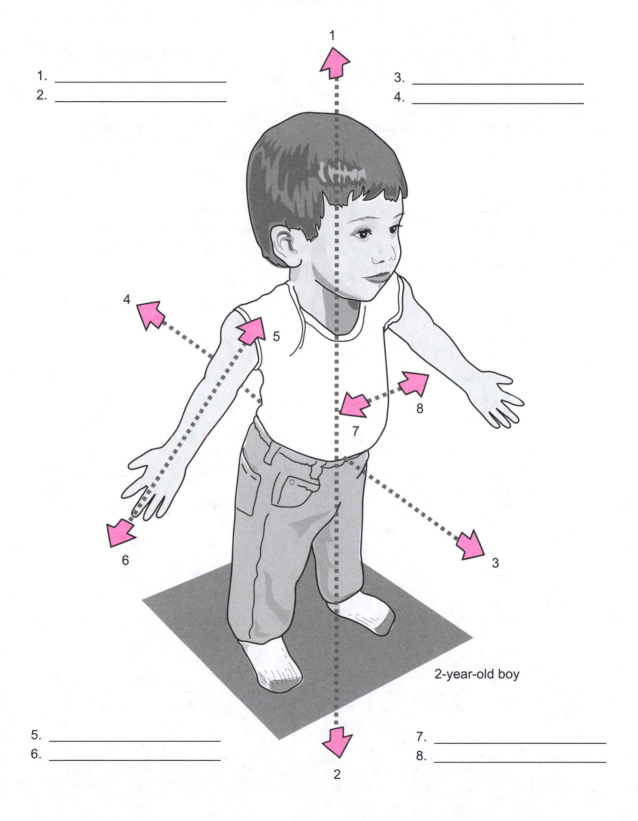

1. _____
2. _____

3. _____
4. _____

2-year-old boy

5. _____
6. _____

7. _____
8. _____

Description   Regional terms are part of the working language of anatomy. Many anatomical structures are named after the region of the body in which they are found. For example, there is the *brachial* artery, the *femoral* nerve, and the *deltoid* muscle. Memorizing these terms now will serve you well throughout your study of anatomy.

    The table below gives the key to the numbers on the illustration as well as the definition of each regional term.

Regional Term	Description	Regional Term	Description
1. Cephalic	Head	22. Inguinal	Groin
2. Frontal	Forehead	23. Coxal	Hip
3. Orbital	Eye	24. Pubic	Anterior region of the pelvis
4. Nasal	Nose	25. Femoral	Thigh
5. Buccal	Cheek	26. Patellar	Knee
6. Oral	Mouth	27. Crural	Leg
7. Mental	Chin	28. Tarsal	Ankle
8. Cervical	Neck	29. Pes	Foot
9. Deltoid	Shoulder	30. Cranial	Skull
10. Pectoral	Chest	31. Occipital	Back of head
11. Sternal	Sternum	32. Otic	Ear
12. Axillary	Armpit	33. Thoracic	Chest (*thorax*)
13. Mammary	Breast	34. Vertebral	Spinal column
14. Brachial	Arm	35. Lumbar	Lower back
15. Antecubital	Front of arm	36. Olecranon	Elbow
16. Abdominal	Abdomen	37. Gluteal	Buttock
17. Antebrachial	Forearm	38. Manus	Hand
18. Carpal	Wrist	39. Perineal	Region between the anus and the genitals
19. Palmar	Palm		
20. Digital	Finger	40. Popliteal	Back of knee
21. Pelvic	Pelvis	41. Calcaneal	Heel

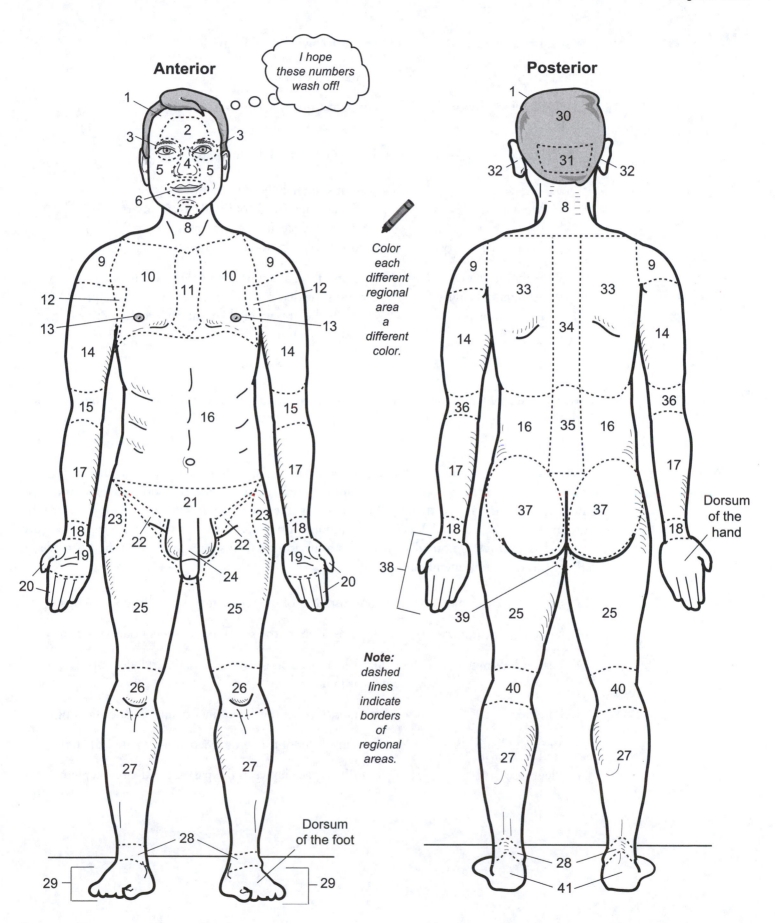

Body Cavities

1. **Dorsal body cavity**—consists of the cranial and vertebral cavities

2. Cranial cavity	contains the brain
3. Vertebral cavity	contains the spinal cord

4. **Ventral body cavity**—consists of the thoracic and abdominopelvic cavities

5. Thoracic cavity	contains the heart and the lungs
6. Mediastinum	median compartment of the thoracic cavity
7. Pleural cavity	fluid-filled space around the lungs
8. Pericardial cavity	fluid-filled space around the heart
9. Abdominopelvic cavity	consists of the abdominal and pelvic cavities
10. Abdominal cavity	contains the digestive organs, kidneys, and ureters
11. Pelvic cavity	contains the urinary bladder, internal reproductive organs, and the rectum

Membranes

Serous membranes are double-layered, fluid-filled sacs that surround organs like the heart and the lungs.

Analogy

To visualize how a serous membrane surrounds an organ, imagine a fist pushed into a partially inflated balloon. The fist is like the **organ** and the balloon is like the **serous membrane**. The **inner layer** of the balloon that touches the fist is like the **visceral layer**. The **outer wall** of the balloon is like the **parietal layer**. The balloon's **inner space** that is filled with air is like the **serous cavity** which is normally filled with a lubricant called serous fluid. This fluid is made by cells within the serous membrane.

12. Parietal pericardium	outermost layer of the serous membrane around the heart
13. Visceral pericardium	innermost layer of the serous membrane around the heart
14. Pericardial cavity	fluid-filled space between the parietal and visceral pericardium
15. Parietal pleura	outermost layer of the serous membrane around the lungs
16. Visceral pleura	innermost layer of the serous membrane around the lungs
17. Pleural cavity	fluid-filled space between the parietal and visceral pleura

BODY CAVITIES

I have this EMPTY feeling inside

Color the body cavities different colors.

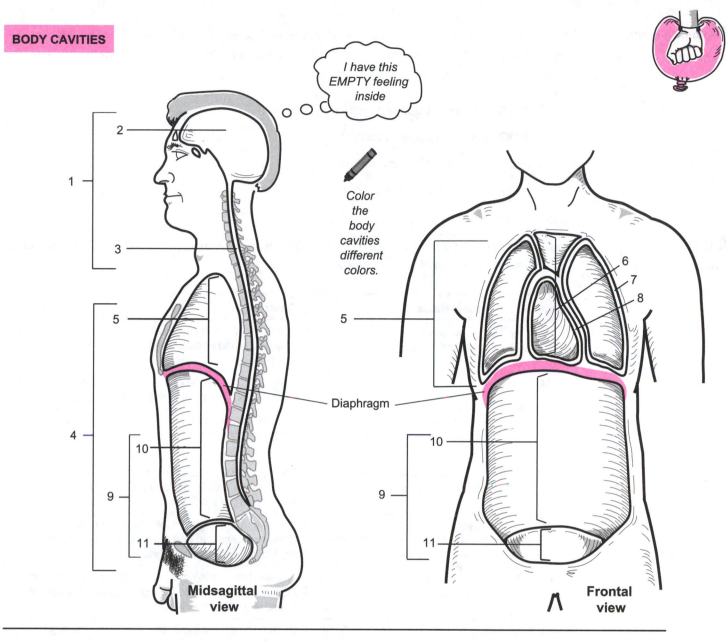

Diaphragm

Midsagittal view

Frontal view

MEMBRANES

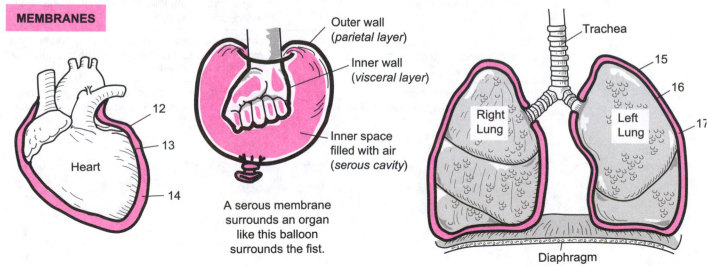

Outer wall (*parietal layer*)

Inner wall (*visceral layer*)

Inner space filled with air (*serous cavity*)

Heart

A serous membrane surrounds an organ like this balloon surrounds the fist.

Trachea

Right Lung

Left Lung

Diaphragm

Abdominopelvic quadrants

Medical professionals often divide the **abdominopelvic region** into the following four areas to help locate pain sites due to injuries and other medical problems. Note that the terms "left" and "right" refers to the subject's left and right.

RUQ	=	Right upper quadrant
LUQ	=	Left upper quadrant
RLQ	=	Right lower quadrant
LLQ	=	Left lower quadrant

Abdominopelvic regions

Anatomists often prefer to divide the abdominopelvic region into a more detailed nine square grid. The table below explains each region.

Symbol	Regional Name	Description
E	Epigastric region (*epi* = above; *gastri* = belly)	Located at the top of the middle column; contains the duodenum and parts of the following: liver, stomach and pancreas.
U	Umbilical region (navel)	Located in the center of the grid; contains parts of both the transverse colon and the small intestine.
H	Hypogastric region (*hypo* = below; *gastri* = belly)	Located at the bottom of the middle column; contains the urinary bladder, sigmoid colon, and part of the small intestine.
RH	Right hypochondriac region (*hypo* = below; *chondro* = cartilage)	Located to the subject's right of the epigastric region; contains the gallbladder and parts of both the right kidney and the liver.
LH	Left hypochondriac region (*hypo* = below; *chondro* = cartilage)	Located to the subject's left of the epigastric region; contains the spleen and parts of the following: stomach, left kidney, and large intestine.
RL	Right lumbar region (*lumbus* = loin)	Located to the subject's right of the umbilical region; contains parts of the following: large intestine, small intestine, and right kidney.
LL	Left lumbar region (*lumbus* = loin)	Located to the subject's left of the umbilical region; contains parts of the following: large intestine, small intestine, and left kidney.
RI	Right iliac region (*iliac* = largest part of hip bone)	Located to the subject's right of the hypogastric region; contains the bottom of the cecum, the appendix, and part of the small intestine
LI	Left iliac region (*iliac* = largest part of hip bone)	Located to the subject's left of the hypogastric region; contains parts of both the large and small intestine

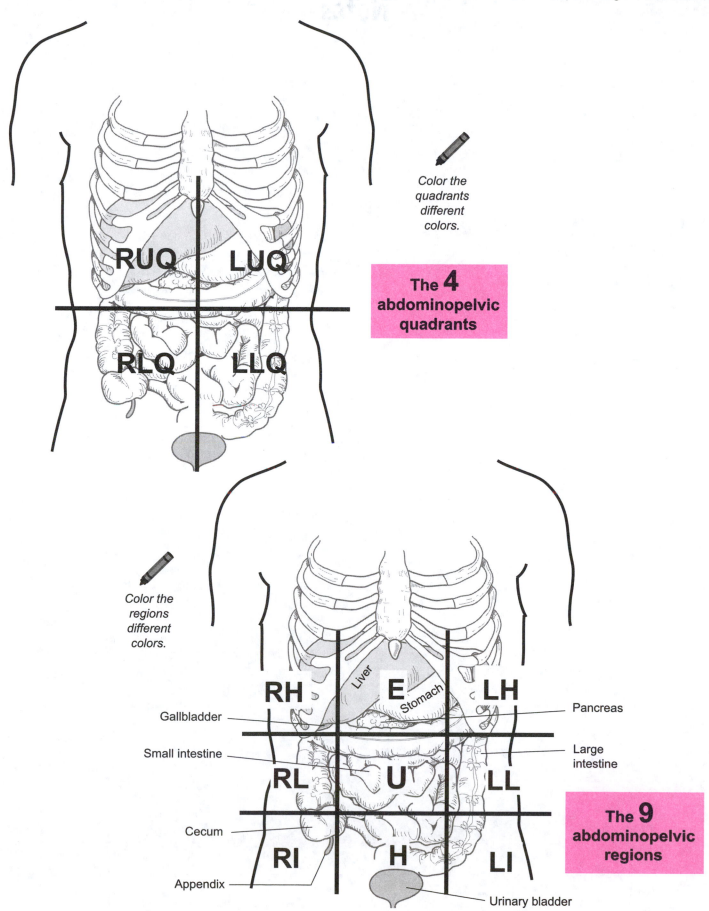

Color the quadrants different colors.

The 4 abdominopelvic quadrants

RUQ LUQ

RLQ LLQ

Color the regions different colors.

The 9 abdominopelvic regions

RH E LH

Liver

Stomach

Gallbladder

Pancreas

Small intestine

Large intestine

RL U LL

Cecum

RI H LI

Appendix

Urinary bladder

Notes

Cells

Description

The cell is the basic unit of life. The human organism begins as a single cell called a **zygote** (*fertilized egg*). Then it divides over and over again to produce more than a trillion cells in the adult. Though there are many different types of cells with a variety of sizes and shapes, all share basic structural features in common. Each cell is surrounded by a **plasma** (*cell*) **membrane** and contains many different types of organelles (*small organ*), each with its own special function(s).

NONMEMBRANOUS ORGANELLES

Organelle	Function
Centrosome (composed of two centrioles)	Forms poles in cell for movement of chromosomes in cell division.
Cilia	**Plasma membrane** extensions containing microtubules that move materials over the cell.
Microfilament	Connects **cytoskeleton** to **cell membrane**; contraction allows for movement of part of a cell or a change in cell shape.
Microtubules	Hollow tubes of protein which can act as tracks along which organelles move.
Microvilli	Plasma membrane extensions that assist in absorption of nutrients and other substances.
Ribosomes	Composed of two subunits, functions in protein synthesis.

MEMBRANOUS ORGANELLES

Organelle	Function

Endoplasmic reticulum (ER)—(two types)
    Smooth (has no ribosomes attached)
    Rough (has ribosomes attached)
	Smooth: lipid and carbohydrate synthesis
	Rough: modification and packaging of newly made proteins

Golgi complex (Golgi body; Golgi apparatus)	Finishes, stores, distributes chemical products of the cell (*e.g.,* proteins)
Lysosome	Vesicles containing digestive enzymes to remove pathogens and broken organelles
Mitochondrion	Site of aerobic cellular respiration; produces ATP for the cell
Nucleus • nucleolus (region within nucleus containing DNA and RNA)	Large structure that contains DNA—the genetic material for making proteins
Peroxisome	Vesicles containing enzymes to break down substances such as hydrogen peroxide, fatty acids, amino acids; detoxifies toxic substances

Key to Illustration

1. Microvilli
2. Plasma *(cell)* membrane
3. Smooth endoplasmic reticulum
4. Nuclear pores
5. Peroxisome
6. Nucleus
7. Nucleoplasm
8. Nucleolous
9. Ribosomes
10. Rough endoplasmic reticulum
11. Cytoskeleton
12. Mitochondrion
13. Lysosome
14. Vacuole
15. Microfilament
16. Golgi complex
17. Cytoplasm
18. Cilia
19. Microtubule
20. Centrioles

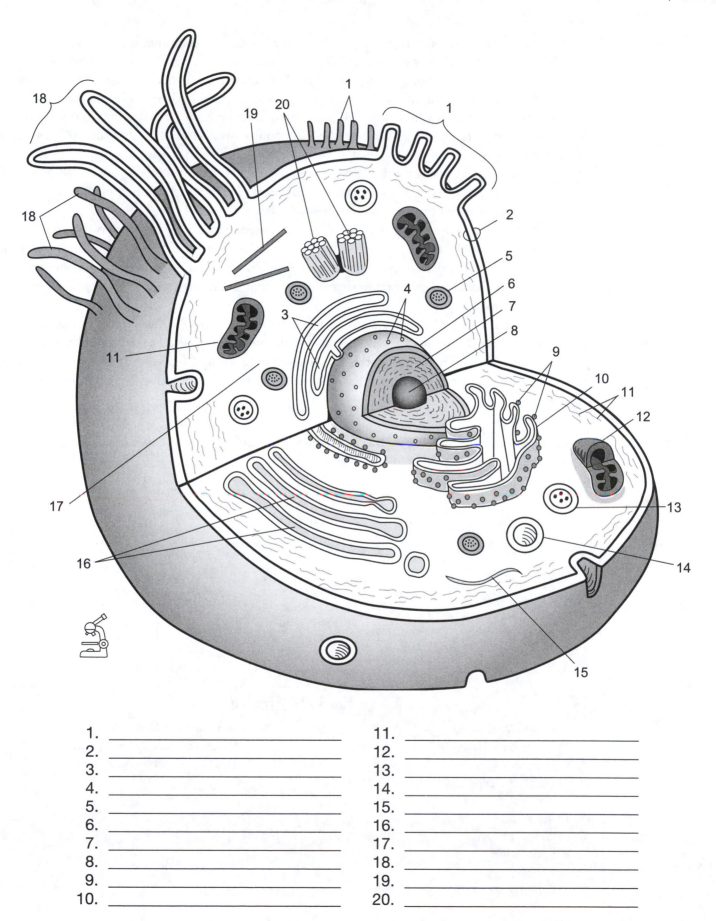

1. _____
2. _____
3. _____
4. _____
5. _____
6. _____
7. _____
8. _____
9. _____
10. _____

11. _____
12. _____
13. _____
14. _____
15. _____
16. _____
17. _____
18. _____
19. _____
20. _____

Description

Every cell is enclosed by an envelope called a **plasma** *(cell)* **membrane**. It is the gateway through which substances enter or exit any cell. Because this structure is the first level of interaction with any cell, it is very important in physiology. The fundamental repeating unit within a plasma membrane is a **phospholipid molecule** that has two parts: a spherical **headgroup** and a **tailgroup** consisting of two fatty acid chains. The headgroup is hydrophilic *(water-loving)* so it is chemically attracted to water molecules, while the tailgroup is hydrophobic *(water-fearing)*, so it is not attracted to water molecules.

Phospholipid molecules align and cluster together to form layers. A single layer is called a **phospholipid monolayer**. Because plasma membranes have two of these layers, the entire membrane is generally referred to as a **phospholipid bilayer**.

Proteins are scattered throughout the plasma membrane. Those that span the entire bilayer are called **integral proteins** and they have a variety of functions. Some act as special channels through which only certain types of ions can pass. Others act as receptors for hormones or neurotransmitters. Some integral proteins have polysaccharide chains attached to them and are called **glycoproteins**.

Peripheral proteins are a separate category because they connect to only one surface of the membrane. Some may act as enzymes, while others may serve as a structural component of the cytoskeleton. The **cytoskeleton** is composed of a series of filaments and is located beneath the phospholipid bilayer. It serves as a kind of structural scaffolding that supports the plasma membrane.

Function

Selectively permeable membrane; it controls what can enter and exit the cell based on factors such as size, charge, and lipid solubility.

Key to Illustration

Plasma (Cell) Membrane

1. Polysaccharide chain
2. Glycoprotein
3. Phospholipid monolayer
4. Cholesterol molecule
5. Filaments of cytoskeleton
6. Peripheral protein
7. Integral protein
8. Glycolipid
9. Phospholipid bilayer

Phospholipid Molecule

10. Headgroup of phospholipid
11. Tailgroup of phospholipid

Plasma *(cell)* Membrane Structure

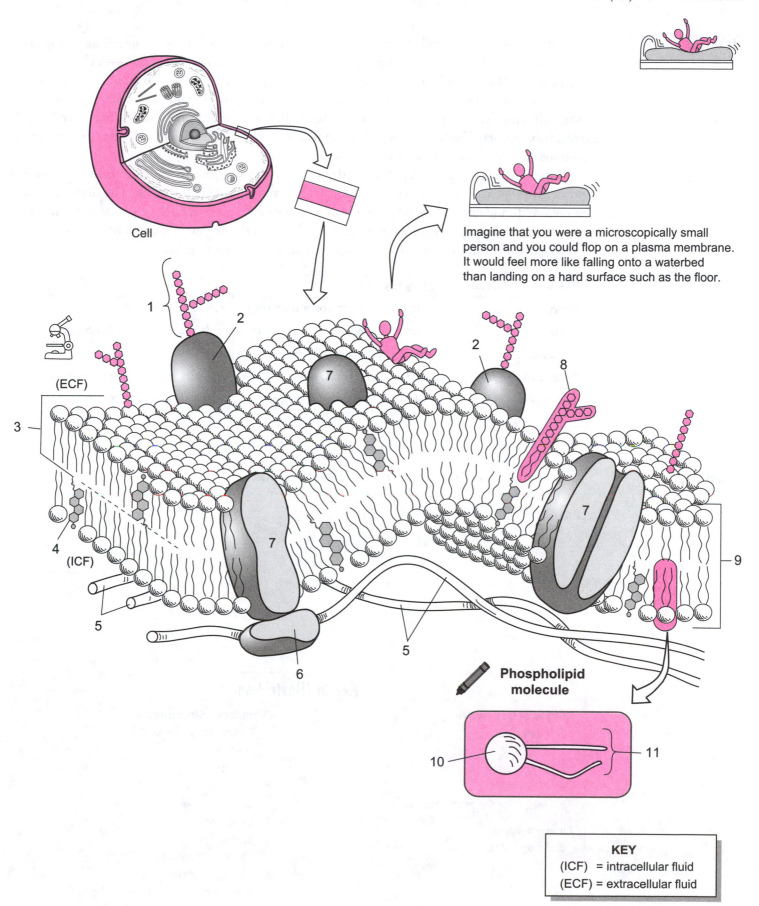

Cell

Imagine that you were a microscopically small person and you could flop on a plasma membrane. It would feel more like falling onto a waterbed than landing on a hard surface such as the floor.

(ECF)

(ICF)

Phospholipid molecule

Description

The life cycle of a cell, called the **cell cycle**, is divided into five phases: **interphase**, **prophase**, **metaphase**, **anaphase**, and **telophase**. During the longest phase—**interphase**—cells grow, develop, and make copies of their chromosomes. Interphase is subdivided into three subphases—G_1, S, and G_2. In G_1 (**Gap 1**)—when the cell is actively growing and synthesizing proteins. For most cells this period lasts from a few minutes to hours. In the S phase (S for **S**ynthetic) DNA replication occurs. The final phase, G_2 (**Gap 2**) is relatively short in duration. Enzymes and other proteins needed for cell division are produced.

After interphase, cells prepare to undergo a cell division process called **mitosis**. Mitosis begins with **prophase** and ends with the formation of two new **daughter cells**. The division of the cytoplasm during mitosis, called **cytokinesis**, can begin in late anaphase and ends after mitosis is completed. Each daughter cell is an exact cloned copy of the original parent cell.

A few key events to identify each stage of mitosis are listed as follows:

Phase of Cell Cycle	Key Event
Interphase	Cell growth; **chromosomes** replicate; protein synthesis
Prophase	Nuclear membrane disappears; **chromatids** visible; **spindle fibers** appear
Metaphase	Sister **chromatids** align along the equator of the cell
Anaphase	**Chromatids** are separated and pulled to opposite poles by **spindle fibers**
Telophase	Cleavage furrow forms; cell pinches in two and divides **cytoplasm**; nuclear membrane reappears

Study Tip

To recall the phases of the cell cycle, use this mnemonic:

IPMAT = "**I** **P**assed **M**y **A**natomy **T**est"

Key to Illustration

Stages of Cell Cycle

1. Interphase
2. Prophase
3. Metaphase
4. Anaphase
5. Telophase *(and cytokinesis)*

Final Products

6. Daughter cells

Significant Structures

a. Plasma *(cell)* membrane
b. Nucleolus
c. Nucleus
d. Centriole
e. Chromosome
f. Centromere
g. Spindle fibers
h. Sister chromatids
i. Cleavage furrow
j. Nuclear membrane *(still forming)*

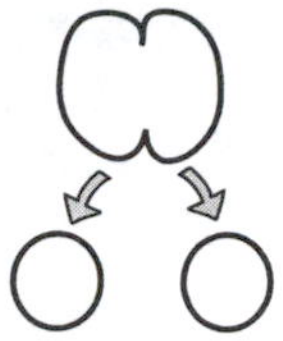

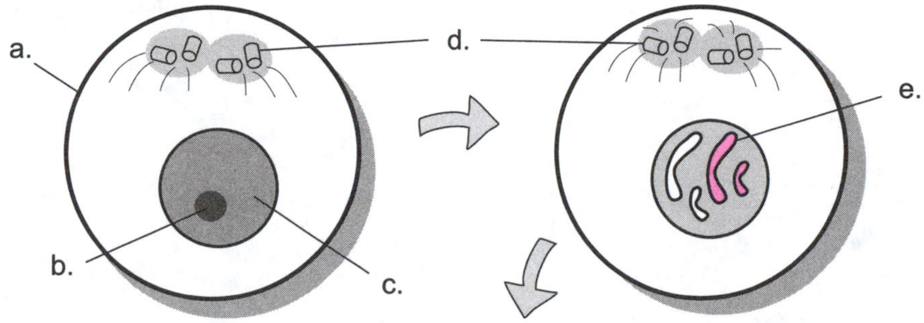

a.

b.

c.

d.

e.

1. _____

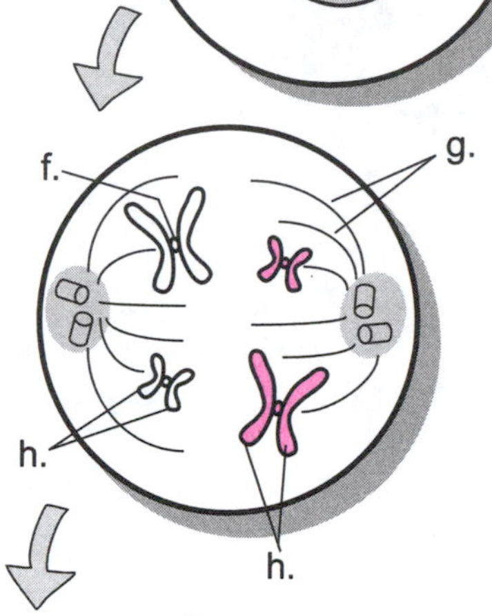

f.

g.

h.

h.

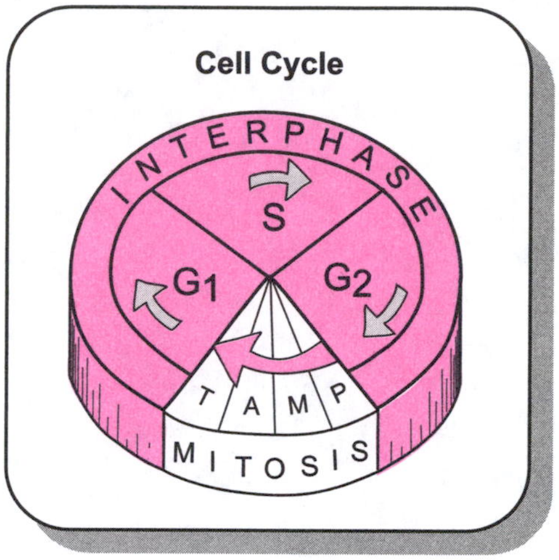

Cell Cycle

INTERPHASE

S

G1

G2

T A M P

MITOSIS

To recall the
phases of the
cell cycle, use
this mnemonic:

I
*P*assed
*M*y
*A*natomy
*T*est

2. _____

i.

KEY

P = prophase

M = metaphase

A = anaphase

T = telophase

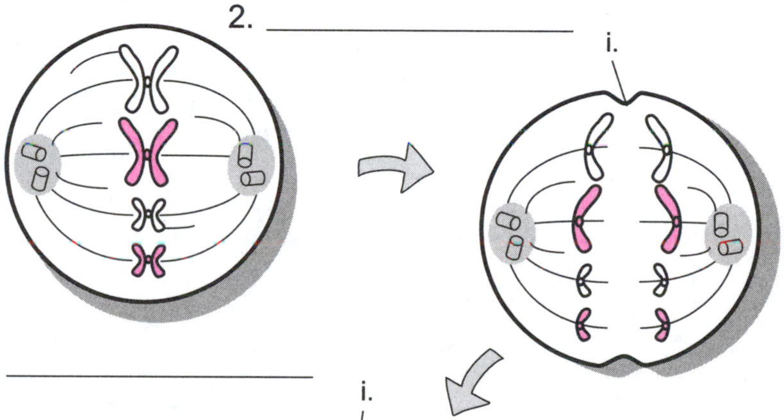

3. _____

i.

4. _____

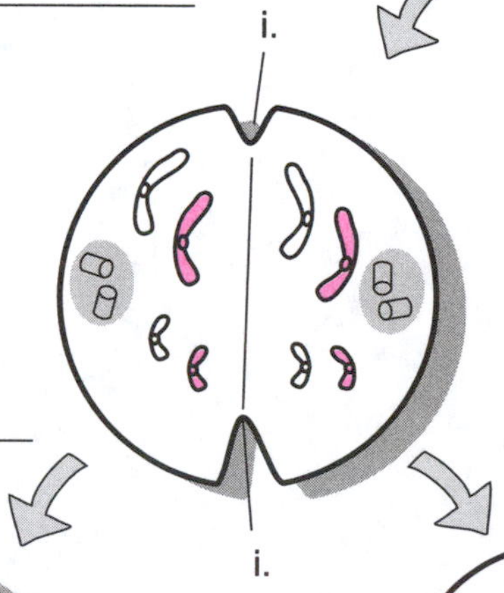

i.

5. _____

i.

j.

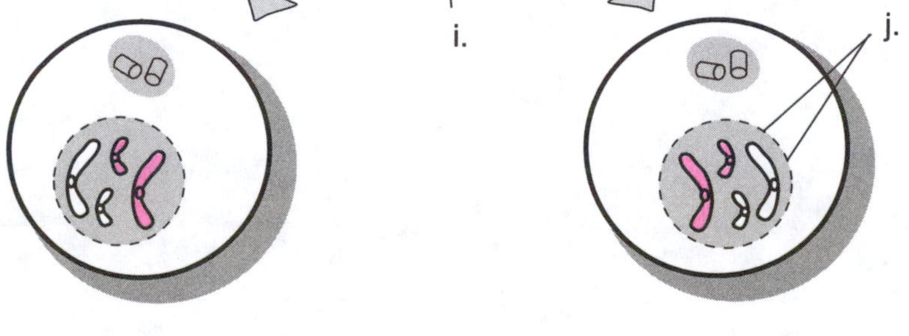

6. _____

6. _____

Notes

Tissues

Description

Epithelial tissues line internal cavities and passageways and cover external body surfaces. They are composed mostly of cells that rest on a thin **basement membrane**. No blood vessels are present. One method of classifying epithelial tissues is by the number of layers of cells. *Simple* epithelia have a single layer of cells, and *stratified* epithelia have multiple layers of cells. Classification is also based on the following cell shapes: *squamous* (thin, flat), *cuboidal* (cube-shaped), and *columnar* (column-shaped).

Simple squamous epithelium is a single row of thin, flat cells.

Analogy

Each **simple squamous epithelial cell** is compared to a **fried egg** because both are flat with an irregular border. The yolk is like the nucleus of the cell.

Location

Lines internal surface of ventral body cavities, blood vessels, and heart; parts of kidney tubules; alveoli of the lungs.

Function

Flat shape allows substances to either diffuse easily through the cell or be filtered through it; secretion; reduces friction.

Key to Illustration

1. Individual simple squamous epithelial cell
2. Nucleus of simple squamous epithelial cell
3. Alveoli

Location

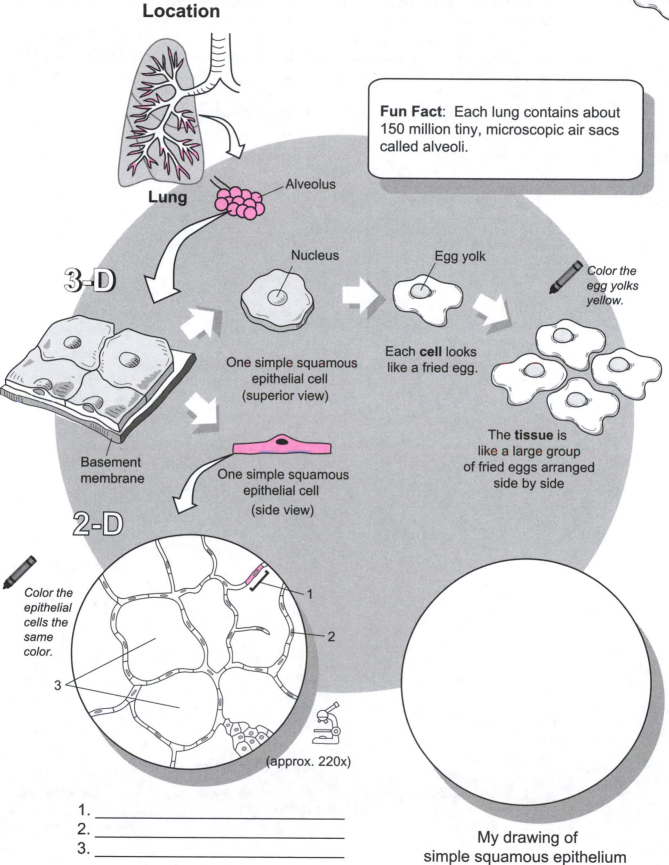

Lung

Alveolus

Fun Fact: Each lung contains about 150 million tiny, microscopic air sacs called alveoli.

3-D

Nucleus

Egg yolk

Color the egg yolks yellow.

One simple squamous epithelial cell (superior view)

Each **cell** looks like a fried egg.

The **tissue** is like a large group of fried eggs arranged side by side

Basement membrane

One simple squamous epithelial cell (side view)

2-D

Color the epithelial cells the same color.

1

2

3

(approx. 220x)

1. _____
2. _____
3. _____

My drawing of simple squamous epithelium

29

Description

Epithelial tissues line internal cavities and passageways and cover external body surfaces. They are composed mostly of cells that rest on a thin **basement membrane**. No blood vessels are present. One method of classifying epithelial tissues is by the number of layers of cells. *Simple* epithelia have a single layer of cells, and *stratified* epithelia have multiple layers of cells. Classification is also based on the following cell shapes: *squamous* (thin, flat), *cuboidal* (cube-shaped), and *columnar* (column-shaped).

    Simple cuboidal epithelium is a single row of cube-shaped cells.

Analogy

Each **simple cuboidal cell** is shaped like an **ice cube**.

Location

Ducts of glands, parts of kidney tubules; follicles of thyroid gland

Function

Secretion; absorption

Key to Illustration

1. Simple cuboidal epithelial cell
2. Nucleus of simple cuboidal epithelial cell
3. Connective tissue

Location

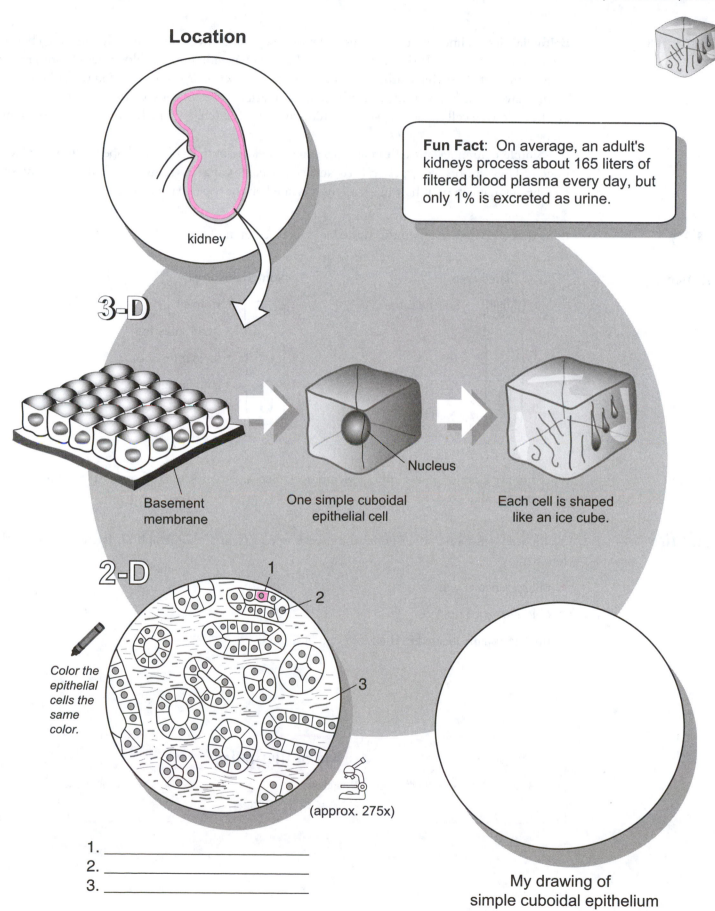

kidney

Fun Fact: On average, an adult's kidneys process about 165 liters of filtered blood plasma every day, but only 1% is excreted as urine.

3-D

Basement membrane

One simple cuboidal epithelial cell

Nucleus

Each cell is shaped like an ice cube.

2-D

Color the epithelial cells the same color.

1
2
3

(approx. 275x)

1. _____
2. _____
3. _____

My drawing of simple cuboidal epithelium

Description

Epithelial tissues line internal cavities and passageways and cover external body surfaces. They are composed mostly of cells that rest on a thin **basement membrane**. No blood vessels are present. One method of classifying epithelial tissues is by the number of layers of cells. *Simple* epithelia have a single layer of cells, and *stratified* epithelia have multiple layers of cells. Classification is also based on the following cell shapes: *squamous* (thin, flat), *cuboidal* (cube-shaped), and *columnar* (column-shaped).

    Simple columnar epithelium appears as a single layer of tall, column-shaped cells with oblong nuclei. They are of two types: *ciliated* and *non-ciliated*. **Cilia** are numerous folds in the **plasma membrane** that appear as hair-like structures located on the top of each cell.

Analogy

Each simple columnar cell in this tissue looks like a column.

Location

Ciliated type

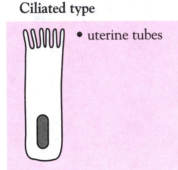

• uterine tubes

Non-ciliated type

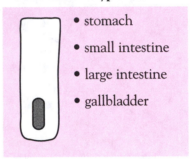

• stomach

• small intestine

• large intestine

• gallbladder

Function

The primary function is absorption; secretion of mucus, enzymes, and other substances; movement of mucus by cilia.

Study Tips

To identify this tissue either under the microscope or from a photograph or diagram, look for the following:

● tall rectangular cells

● oblong-shaped nucleus

● nucleus usually located in the lower half of the cell

Key to Illustration

1. Simple columnar epithelial cell
2. Nucleus
3. Basement membrane
4. Connective tissue

Location

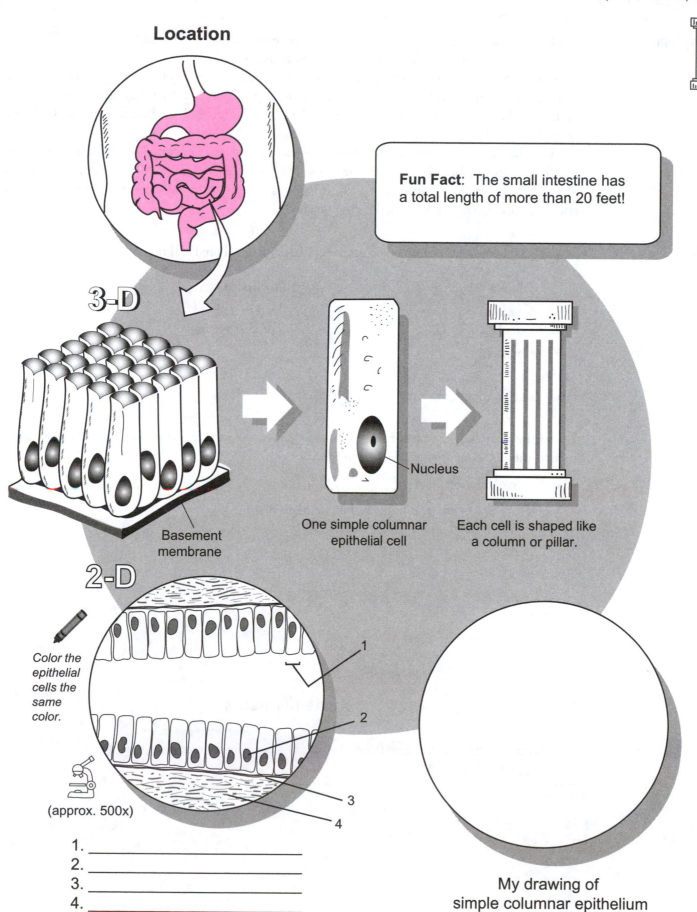

Fun Fact: The small intestine has a total length of more than 20 feet!

3-D

Basement membrane

Nucleus

One simple columnar epithelial cell

Each cell is shaped like a column or pillar.

2-D

Color the epithelial cells the same color.

(approx. 500x)

1. _____
2. _____
3. _____
4. _____

My drawing of simple columnar epithelium

33

Description

Epithelial tissues line internal cavities and passageways and cover external body surfaces. They are composed mostly of cells that rest on a thin **basement membrane**. No blood vessels are present. One method of classifying epithelial tissues is by the number of layers of cells. *Simple* epithelia have a single layer of cells, and *stratified* epithelia have multiple layers of cells. Classification is also based on the following cell shapes: *squamous* (thin, flat), *cuboidal* (cube-shaped), and *columnar* (column-shaped).

Pseudostratified columnar epithelium consists of a single row of cells. Most cells have a columnar shape, while other, shorter cells may look like cuboidal. The term "pseudostratified" literally means "falsely stratified." In other words, it looks as if it has multiple layers but actually has only one layer, because the cells are of differing heights.

Location

The two types of pseudostratified columnar epithelium are ciliated and non-ciliated.

Ciliated
- nasal cavity
- trachea
- bronchi

Non-ciliated
- ducts of male reproductive tract

Function

Protection; secretion; movement of mucus by **cilia**

Study Tips

Under the microscope at higher magnifications, you can use the following landmarks to distinguish pseudostratified columnar epithelium cells:

- Cells have differing heights.

- Nuclei are not in an organized row, but are more staggered.

Key to Illustration

1. Cilia
2. Nucleus of one pseudostratified ciliated columnar epithelial cell

3. Basement membrane
4. Connective tissue

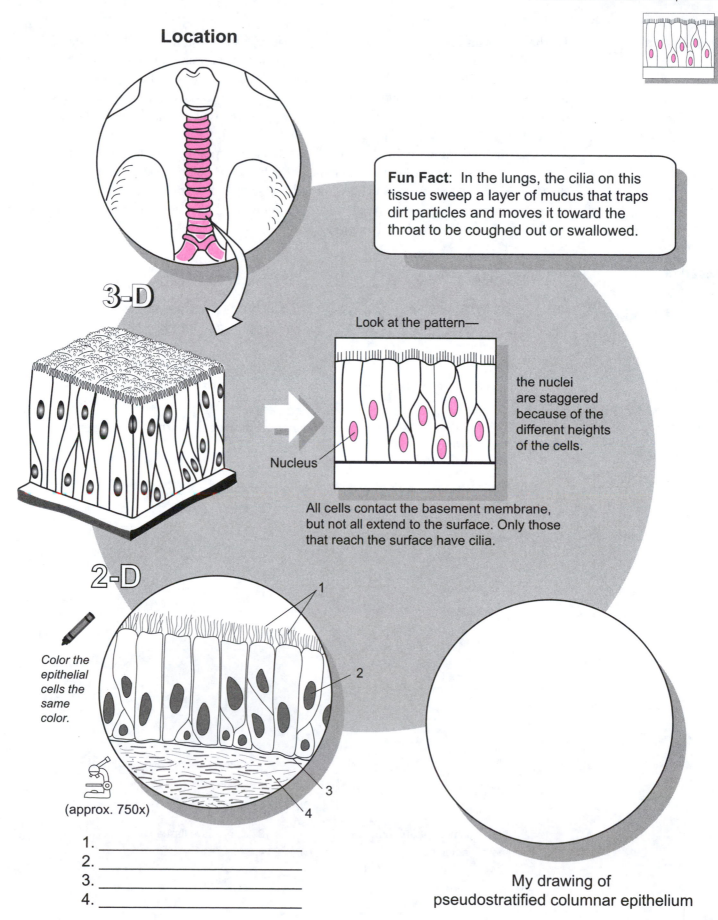

Location

Fun Fact: In the lungs, the cilia on this tissue sweep a layer of mucus that traps dirt particles and moves it toward the throat to be coughed out or swallowed.

3-D

Look at the pattern—

the nuclei are staggered because of the different heights of the cells.

Nucleus

All cells contact the basement membrane, but not all extend to the surface. Only those that reach the surface have cilia.

2-D

1

2

Color the epithelial cells the same color.

3

4

(approx. 750x)

1. _____
2. _____
3. _____
4. _____

My drawing of
pseudostratified columnar epithelium

Description

Epithelial tissues line internal cavities and passageways and cover external body surfaces. They are composed mostly of cells that rest on a thin **basement membrane**. No blood vessels are present. One method of classifying epithelial tissues is by the number of layers of cells. *Simple* epithelia have a single layer of cells, and *stratified* epithelia have multiple layers of cells. Classification is also based on the following cell shapes: *squamous* (thin, flat), *cuboidal* (cube-shaped), and *columnar* (column-shaped).

Stratified squamous epithelium is of two different types—keratinized and non-keratinized. The bottom layer in either type is composed of cuboidal or columnar cells that are active in cell division. New cells are pushed upward toward the surface. In a process called **keratinization** the new cells in the keratinized type fill with a protein called **keratin**. The result is that the outer surface of this tissue is tough and water-resistant.

Location

Keratinized
- epidermis of skin

Non-keratinized
- lining of mouth, pharynx, esophagus, anus, and vagina

Function

Provides physical protection against abrasion and pathogens to underlying tissues.

Study Tip

Pattern: Transition in cell shape from cuboidal or columnar cells in the lower region to flat cells on the top.

Key to Illustration

1. Stratified squamous epithelium
2. Basement membrane
3. Connective tissue
4. Nuclei

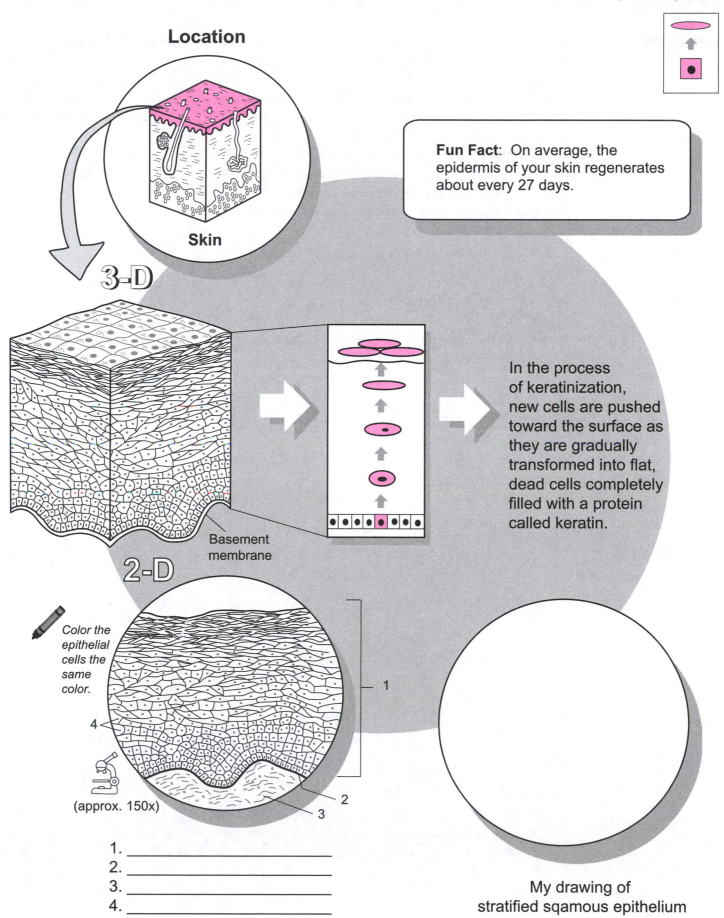

Location

Skin

3-D

Fun Fact: On average, the epidermis of your skin regenerates about every 27 days.

2-D

Basement membrane

In the process of keratinization, new cells are pushed toward the surface as they are gradually transformed into flat, dead cells completely filled with a protein called keratin.

Color the epithelial cells the same color.

(approx. 150x)

1. _____
2. _____
3. _____
4. _____

My drawing of stratified sqamous epithelium

Description

Epithelial tissues line internal cavities and passageways and cover external body surfaces. They are composed mostly of cells that rest on a thin **basement membrane**. No blood vessels are present. One method of classifying epithelial tissues is by the number of layers of cells. *Simple* epithelia have a single layer of cells, and *stratified* epithelia have multiple layers of cells. Classification is also based on the following cell shapes: *squamous* (thin, flat), *cuboidal* (cube-shaped), and *columnar* (column-shaped).

Transitional epithelium is able to stretch and recoil so it can be illustrated in either a stretched or a relaxed state. In the relaxed state it appears to be composed of a variety of cell shapes. On the bottom it may contain cuboidal or columnar cells, and cells at the surface are large, dome-shaped cells that transform into a squamous shape when stretched.

Location

Lines ureters, urinary bladder, urethra, and renal pelvis.

Function

Easily allows stretching and recoiling.

Study Tip

Pattern: In the relaxed state, look for the cell pattern of cuboidal cells near the bottom, columnar in the middle, and large, dome-shaped cells on top.

Key to Illustration

1. Stratified transitional epithelium

2. Basement membrane

3. Connective tissue

Location

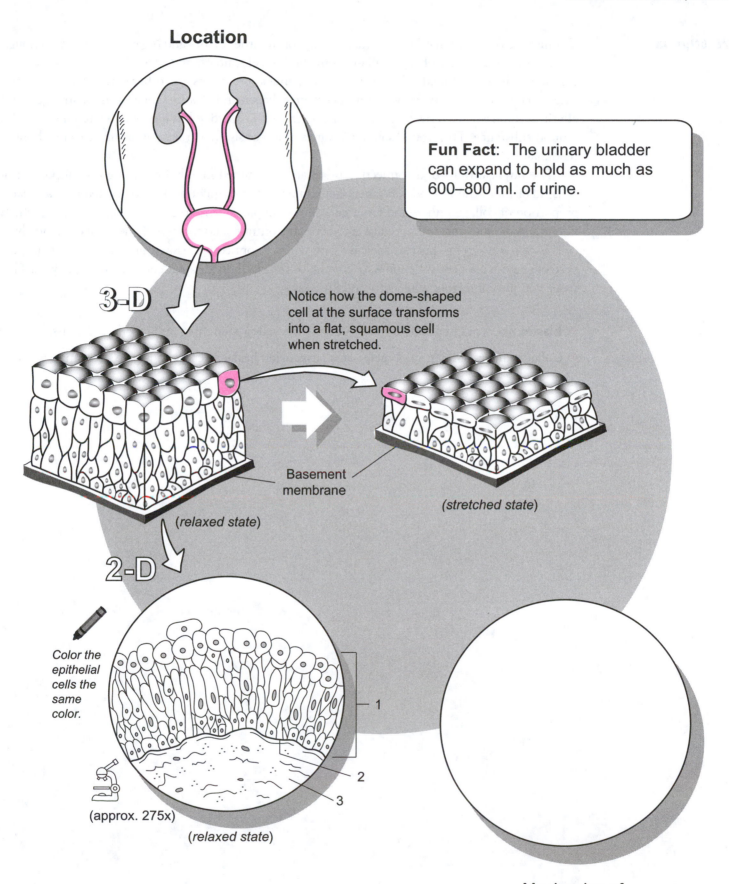

Fun Fact: The urinary bladder can expand to hold as much as 600–800 ml. of urine.

3-D

Notice how the dome-shaped cell at the surface transforms into a flat, squamous cell when stretched.

Basement membrane

(relaxed state)

(stretched state)

2-D

Color the epithelial cells the same color.

1

2

3

(approx. 275x)

(relaxed state)

My drawing of transitional epithelium

Description

Connective tissues primarily give structural support to other tissues and organs in the body. Though there are a wide variety of types, all are composed of cells, fibers, and ground substance. The most common cell type is called a **fibroblast**, which manufactures fibers and other extracellular material. The two most common types of protein fibers are **collagen** and **elastin**. Collagen is for strength, and elastin is for elasticity. The cells and the fibers are both embedded in a gel-like material called the **ground substance**. The ground substance varies in its consistency from gelatin-like to a much more rigid material.

The physical traits of a connective tissue are determined mainly by varying the proportion of cells, fibers, and the ground substance. For example, a strong connective tissue requires a greater proportion of collagen fibers and fewer cells. An example is dense regular connective tissue, which is found in tendons that anchor muscle to bone. In contrast, a connective tissue composed mostly of cells is not very strong. Such is the case with adipose connective tissue (*fat tissue*). The main purpose of adipose connective tissue is to store lipids (*fat*) in individual fat cells called **adipocytes**. This tissue contains numerous adipocytes and little else.

Analogy

- **Elastin fibers** are like **rubber bands** because they allow stretching and recoiling in a tissue.

- **Collagen fibers** are like **steel cables on a suspension bridge** because they give strength to a tissue.

All connective tissues contain the following basic components:

Cells	+	Fibers	+	Ground substance	=

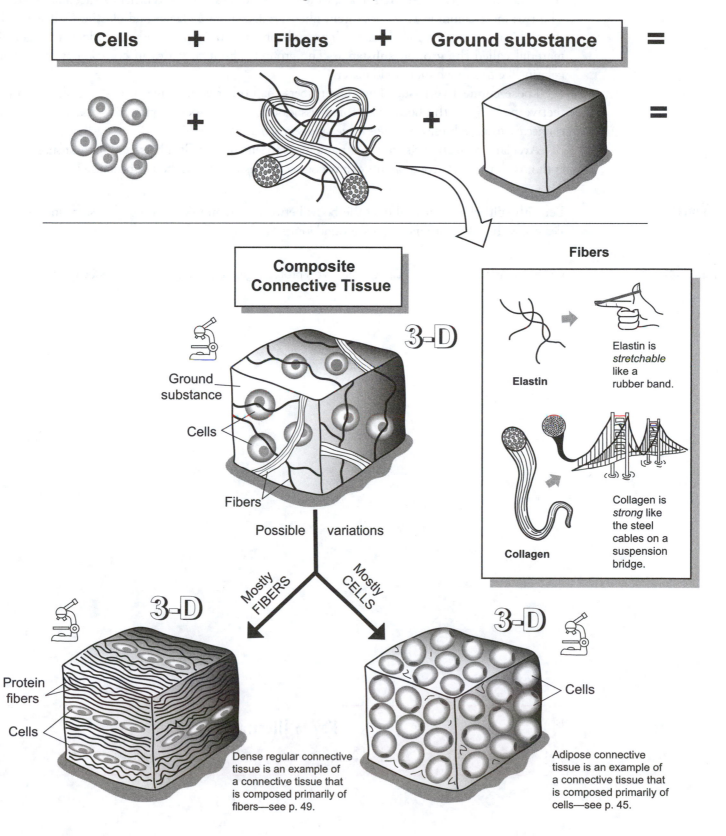

Composite Connective Tissue

3-D

Ground substance

Cells

Fibers

Possible variations

Mostly FIBERS

Mostly CELLS

Fibers

Elastin

Elastin is *stretchable* like a rubber band.

Collagen

Collagen is *strong* like the steel cables on a suspension bridge.

3-D

Protein fibers

Cells

Dense regular connective tissue is an example of a connective tissue that is composed primarily of fibers—see p. 49.

3-D

Cells

Adipose connective tissue is an example of a connective tissue that is composed primarily of cells—see p. 45.

Description

Connective tissues primarily give structural support to other tissues and organs in the body. Though there are a wide variety of types, all are composed of cells, fibers, and ground substance. The most common cell type is called a **fibroblast**, which manufactures fibers and other extracellular material. The two most common types of protein fibers produced are **collagen** and **elastin**. Collagen is for strength and elastin is for elasticity. The cells and the fibers are both embedded in a gel-like material called the **ground substance**. The ground substance varies in consistency from being gelatin-like to a much more rigid material.

Loose connective tissues have fewer fibers than other connective tissues and serve as a protective padding in the body. The three tissues classified as loose connective tissues are: *areolar connective tissue*, *adipose connective tissue*, and *reticular connective tissue*.

Areolar connective tissue has a random arrangement of cells, fibers, and ground substance. It contains all the basic components of any connective tissue without being specialized.

Location

Beneath epithelial tissues all over the body; between skin and skeletal muscles; surrounding blood vessels; within skin; around organs; around joints

Function

Cushions and protects organs; its phagocytes protect against pathogens; holds tissue fluid

Key to Illustration

1. Collagen fibers	2. Elastin fibers	3. Fibroblast nuclei

Location

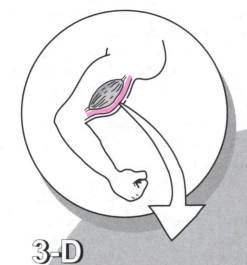

Fun Fact: When hunters skin an animal, the tissue they break to separate skin from muscle is areolar connective tissue.

3-D

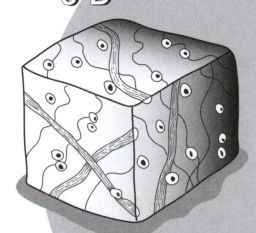

R A N D O M

This tissue type is not specialized, so it does not look like anything in particular. It has a random arrangement of fibers and cells—nothing special!

2-D

Color the cells and fibers different colors.

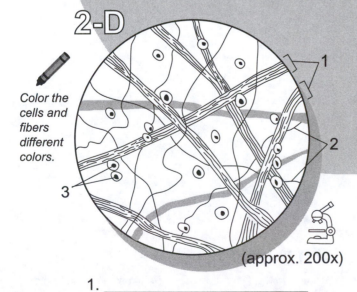

(approx. 200x)

1. _____

2. _____

3. _____

My drawing of
areolar connective tissue

Description

Connective tissues primarily give structural support to other tissues and organs in the body. Though there are a wide variety of types, all are composed of cells, fibers, and ground substance. The most common cell type, called a **fibroblast**, manufactures fibers and other extracellular material. The two most common types of protein fibers produced are **collagen** and **elastin**. Collagen is for strength and elastin is for elasticity. The cells and the fibers are both embedded in a gel-like material called the **ground substance**. The ground substance varies in its consistency from being gelatin-like to a much more rigid material.

Loose connective tissues have fewer fibers than other connective tissues and serve as a protective padding in the body. There are three tissues classified as loose connective tissues: *areolar connective tissue*, *adipose connective tissue*, and *reticular connective tissue*.

Adipose connective tissue is fat tissue. It is composed almost entirely of fat cells called **adipocytes** along with some blood vessels. These cells have a large vacuole to store lipids (*fats*). Though adipocytes are not able to divide, they do change in size by expanding or shrinking depending on the amount of lipid that is stored inside their vacuoles. For example, as a person loses weight, the amount of lipid in the adipocyte's vacuole decreases, causing the cell to shrink in size. Unfortunately, if a person regains that weight, the cells are able to expand back to their original size.

Location

Under all skin but especially in abdomen, buttocks, and breasts; around some organs such as eyeballs and kidneys.

Function

Protects certain organs and other structures; insulates against heat loss through the skin; stores energy as a reserve fuel.

Key to Illustration

1. Blood vessel
2. Nuclei of adipocytes *(fat cells)*
3. Vacuole for lipid storage
4. Plasma membrane of adipocyte *(fat cell)*

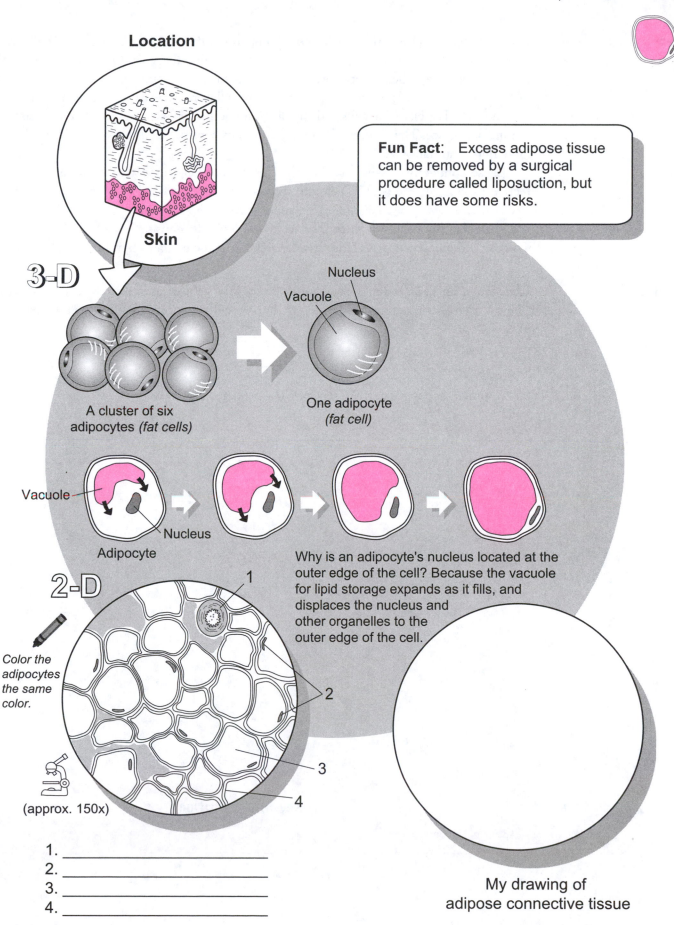

Location

Skin

Fun Fact: Excess adipose tissue can be removed by a surgical procedure called liposuction, but it does have some risks.

3-D

Nucleus

Vacuole

A cluster of six adipocytes *(fat cells)*

One adipocyte *(fat cell)*

Vacuole

Nucleus

Adipocyte

Why is an adipocyte's nucleus located at the outer edge of the cell? Because the vacuole for lipid storage expands as it fills, and displaces the nucleus and other organelles to the outer edge of the cell.

2-D

Color the adipocytes the same color.

1

2

3

4

(approx. 150x)

1. _____
2. _____
3. _____
4. _____

My drawing of adipose connective tissue

Description

Connective tissues primarily give structural support to other tissues and organs in the body. Though there are a wide variety of types, all are composed of cells, fibers, and ground substance. The most common cell type, called a **fibroblast**, manufactures fibers and other extracellular material. The two most common types of protein fibers produced are **collagen** and **elastin**. Collagen is for strength and elastin is for elasticity. The cells and the fibers are both embedded in a gel-like material called the **ground substance**. The ground substance varies in its consistency from being gelatin-like to a much more rigid material.

Loose connective tissues have fewer fibers than other connective tissues and serve as a protective padding in the body. There are three tissues classified as loose connective tissues: *areolar connective tissue*, *adipose connective tissue*, and *reticular connective tissue*.

Reticular (*reticulata* = net) **connective tissue** primarily consists of a network of reticular fibers. The most common cell type is the reticular cell, but it also contains fibroblasts and macrophages.

Analogy

Reticular connective tissue is like **many cobwebs**. The **cobweb** itself is like the **network of reticular fibers** scattered throughout the tissue, which physically supports a variety of cell types.

Location

Spleen, bone marrow, lymph nodes, liver, and kidney

Function

Fibers form a supportive net-like structure for a variety of cell types.

Key to Illustration

1. Reticular fibers

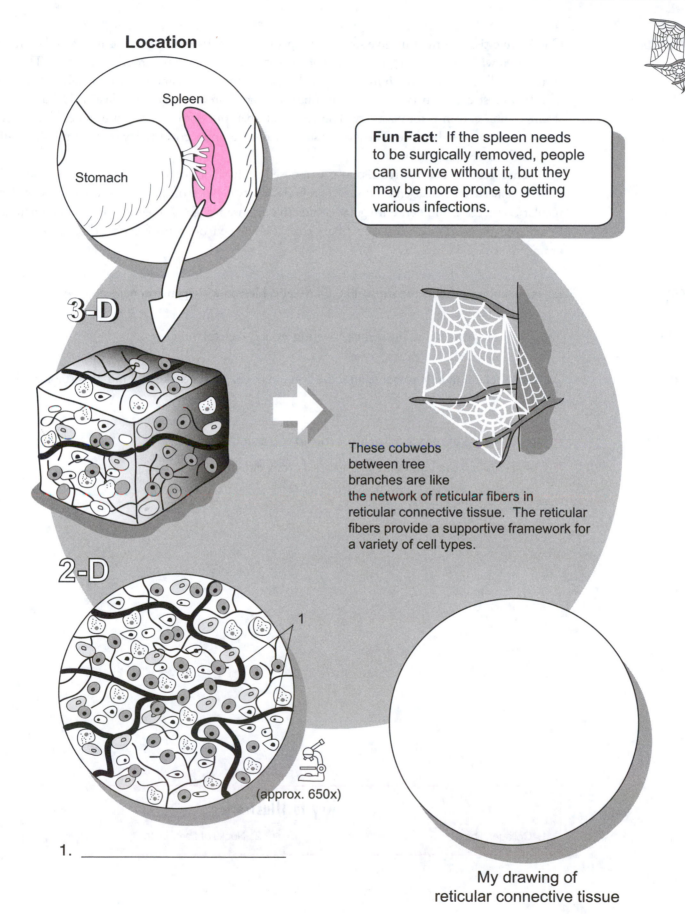

Location

Spleen

Stomach

Fun Fact: If the spleen needs to be surgically removed, people can survive without it, but they may be more prone to getting various infections.

3-D

These cobwebs between tree branches are like the network of reticular fibers in reticular connective tissue. The reticular fibers provide a supportive framework for a variety of cell types.

2-D

1

(approx. 650x)

1. _____

My drawing of reticular connective tissue

Description

Connective tissues primarily give structural support to other tissues and organs in the body. Though there are a wide variety of types, all are composed of cells, fibers, and ground substance. The most common cell type is called a **fibroblast,** which manufactures fibers and other extracellular material. The two most common types of protein fibers produced are **collagen** and **elastin.** Collagen is for strength, and elastin is for elasticity. The cells and fibers are both embedded in a gel-like material called the **ground substance.** The ground substance varies in its consistency from being gelatin-like to a much more rigid material.

Dense regular connective tissue is composed primarily of collagen fibers, so it is also called *fibrous connective* or *collagenous tissue.* The body has two types of dense connective tissue: dense regular connective and dense irregular connective. Dense *regular* connective tissue is characterized by a large proportion of collagen fibers that are stacked on top of each other in an orderly arrangement..

Analogy

Layers of **collagen fibers** are strong like the **steel cables on a suspension bridge.**

Location

Tendons and aponeuroses; ligaments; covering around skeletal muscles.

Function

Anchors skeletal muscle to bone; attaches bone to bone; packages skeletal muscles; stabilizes bones within a joint.

Study Tips

- Fibroblasts are in rows sandwiched between collagen fibers.
- Collagen fibers are layered in an organized arrangement.

Key to Illustration

1. Collagen fibers
2. Nuclei of fibroblasts

Dense *(fibrous)* Regular Connective Tissue

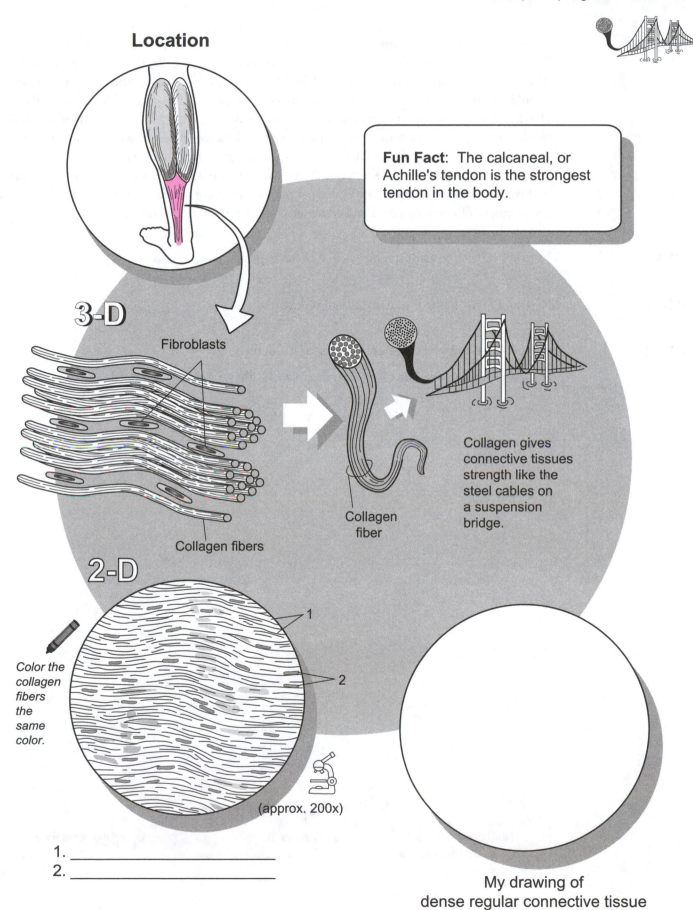

Location

Fun Fact: The calcaneal, or Achille's tendon is the strongest tendon in the body.

3-D

Fibroblasts

Collagen fibers

Collagen fiber

Collagen gives connective tissues strength like the steel cables on a suspension bridge.

2-D

Color the collagen fibers the same color.

1

2

(approx. 200x)

1. _____

2. _____

My drawing of
dense regular connective tissue

Description

Connective tissues primarily give structural support to other tissues and organs in the body. Though they are of a wide variety of types, all are composed of cells, fibers, and ground substance. The most common cell type is called a **fibroblast** which manufactures the fibers and other extracellular material. The two most common types of protein fibers produced are **collagen** and **elastin**. Collagen is for strength and elastin is for elasticity. The cells and the fibers are both embedded in a gel-like material called the **ground substance**. The ground substance varies in its consistency from being almost like gelatin to a much more rigid material.

There are two types of dense connective tissue in the body, namely, dense *regular* connective and dense *irregular* connective. Dense *irregular* connective tissue is characterized by a random arrangement of collagen fibers and a greater proportion of ground substance.

Location

Dermis of the skin; periosteum; visceral organ capsules; around muscles.

Function

Resists stresses applied in many different directions

Study Tips

- Fibroblasts are more scattered throughout the tissue
- Collagen fibers are **not** stacked on top of each other, randomly arranged

Key to Illustration

1. Nucleus of fibroblast	2. Collagen fibers	3. Ground substance

Location

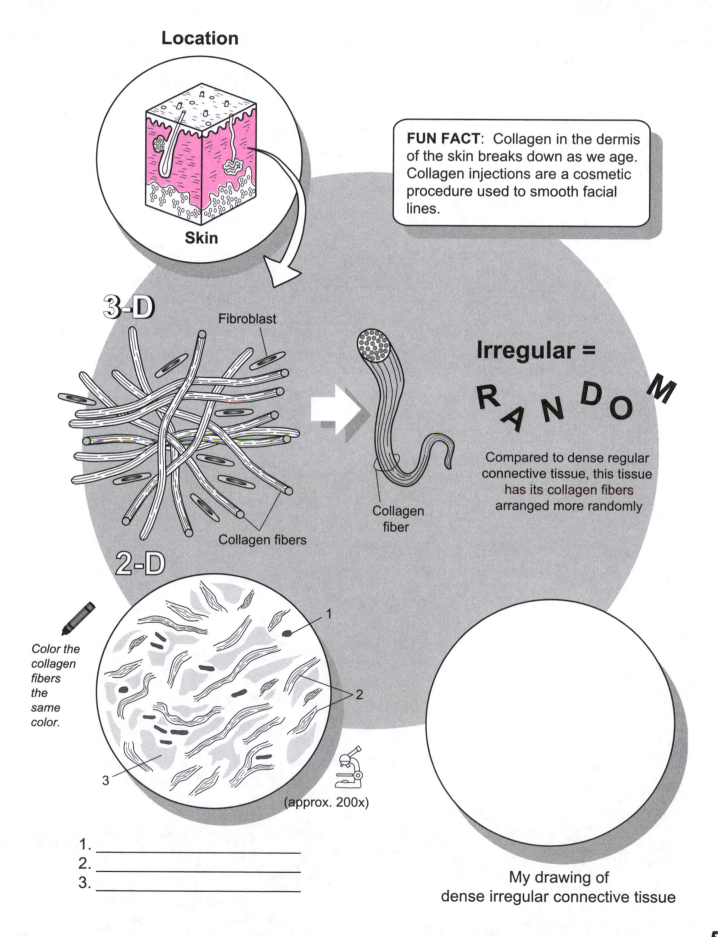

Skin

FUN FACT: Collagen in the dermis of the skin breaks down as we age. Collagen injections are a cosmetic procedure used to smooth facial lines.

3-D

Fibroblast

Collagen fibers

Collagen fiber

Irregular =

R A N D O M

Compared to dense regular connective tissue, this tissue has its collagen fibers arranged more randomly

2-D

Color the collagen fibers the same color.

1

2

3

(approx. 200x)

1. _____

2. _____

3. _____

My drawing of
dense irregular connective tissue

Description

Connective tissues primarily give structural support to other tissues and organs in the body. Though there are a wide variety of types, all are composed of cells, fibers, and matrix.

Cartilage is a specialized type of connective tissue. It is characterized by three traits: **lacunae, chondrocytes,** and a rigid **matrix**. The **matrix** is a firm gel material that contains protein fibers and other substances. Within the matrix are small cavities called **lacunae**. Within the lacunae are living cartilage cells called **chondrocytes**. Because cartilage lacks blood vessels, chondrocytes rely on the diffusion of nutrients into the matrix to survive.

The three basic types of cartilage in the body are:

- **Hyaline cartilage**

- **Elastic cartilage**

- **Fibrocartilage**

Hyaline cartilage is the most common type of cartilage.

Analogy

Three dimensionally, a piece of **any type of cartilage** is similar to a **block of Swiss cheese** in its structure and general consistency. Though cartilage is much stronger, both are solid and flexible. The **cheese** itself is the **matrix** and the **holes** are the **lacunae**.

Location

Covers ends of long bones in synovial joints; between ribs and sternum; cartilages of nose, trachea, larynx, and bronchi; most portions of embryonic skeleton.

Function

Structural reinforcement, slightly flexible support; reduces friction within joints.

Study Tips

To identify this tissue either under the microscope or from a photograph, look for the following:

- This is the **only cartilage type with no apparent fibers** (*they are present but do not stain well*).

- Chondrocytes are evenly scattered within matrix.

Key to Illustration

1. Matrix
2. Lacunae
3. Chondrocytes (*cartilage cells*)
4. Nucleus of a chondrocyte

Location

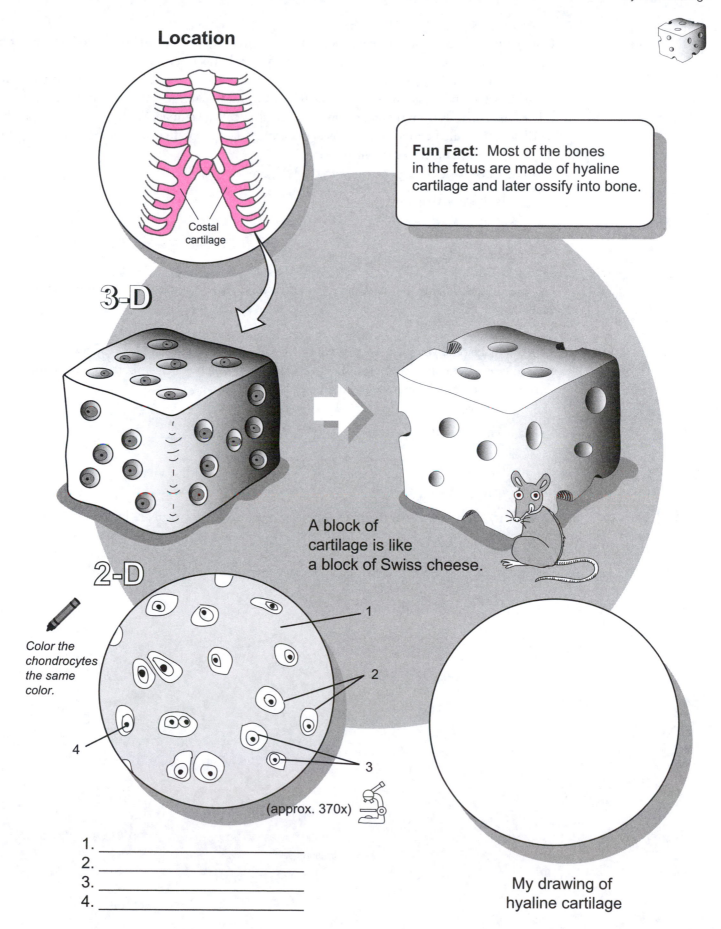

Costal cartilage

Fun Fact: Most of the bones in the fetus are made of hyaline cartilage and later ossify into bone.

3-D

2-D

Color the chondrocytes the same color.

A block of cartilage is like a block of Swiss cheese.

1

2

4

3

(approx. 370x)

1. _____
2. _____
3. _____
4. _____

My drawing of hyaline cartilage

Description

Connective tissues primarily give structural support to other tissues and organs in the body. Though there are a wide variety of types, all are composed of cells, fibers, and matrix.

Cartilage is a specialized type of connective tissue. It is characterized by three traits: **lacunae**, **chondrocytes,** and a rigid **matrix**. The **matrix** is a firm gel material that contains protein fibers and other substances. Within the matrix are small cavities called **lacunae**. Within the lacunae are living cartilage cells called **chondrocytes**. Because cartilage lacks blood vessels, chondrocytes rely on the diffusion of nutrients into the matrix to survive.

The three basic types of cartilage in the body are:

- **Hyaline cartilage**

- **Elastic cartilage**

- **Fibrocartilage**

Elastic cartilage is the most durable and flexible type of cartilage, because of the presence of many elastic fibers.

Analogy

Three dimensionally, a piece of **any type of cartilage** is similar to a **block of Swiss cheese** in its structure and general consistency. Though cartilage is much stronger, both are solid and flexible. The **cheese** itself is the **matrix** and the **holes** are the **lacunae**.

Location

External ear; epiglottis; auditory canal

Function

Provides support while easily returning to original shape when distorted.

Study Tips

To identify this tissue either under the microscope or from a photograph, look for the following:

- Chondrocytes appear larger than other cartilages.

- Numerous elastic fibers have appearance of plant roots branching in the soil.

Key to Illustration

1. Matrix
2. Lacunae
3. Chondrocytes (cartilage cells)
4. Elastin fiber
5. Nucleus of chondrocyte

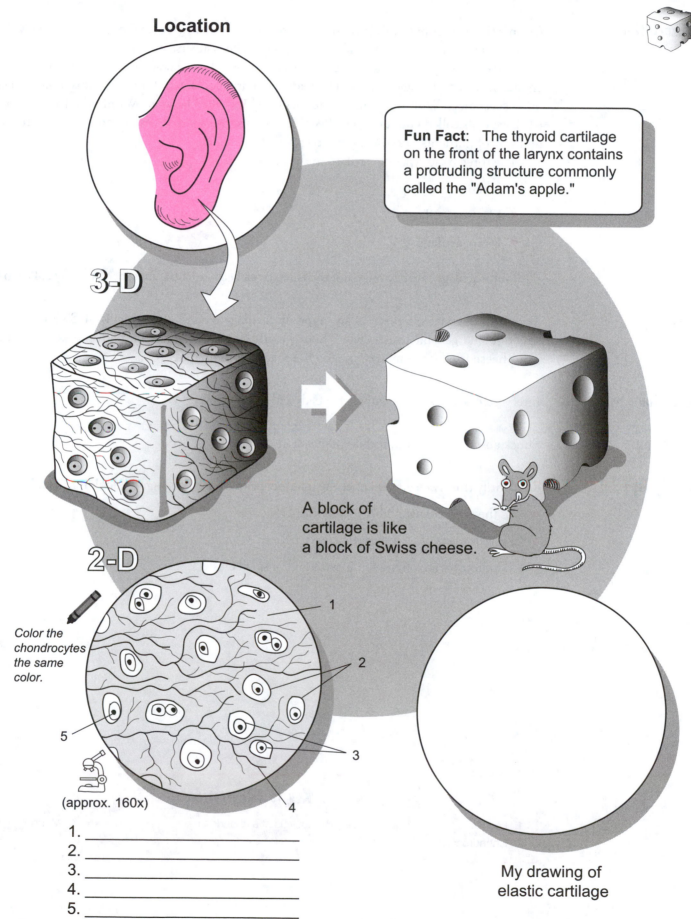

Location

3-D

Fun Fact: The thyroid cartilage on the front of the larynx contains a protruding structure commonly called the "Adam's apple."

A block of cartilage is like a block of Swiss cheese.

2-D

Color the chondrocytes the same color.

(approx. 160x)

My drawing of elastic cartilage

1. _____
2. _____
3. _____
4. _____
5. _____

Description

Connective tissues primarily give structural support to other tissues and organs in the body. Though there are a wide variety of types, all are composed of cells, fibers, and matrix.

Cartilage is a specialized type of connective tissue. It is characterized by three traits: **lacunae**, **chondrocytes,** and a rigid **matrix**. The **matrix** is a firm gel material that contains protein fibers and other substances. Within the matrix are small cavities called **lacunae**. Within the lacunae are living cartilage cells called **chondrocytes**. Because cartilage lacks blood vessels, chondrocytes rely on the diffusion of nutrients into the matrix to survive.

The three basic types of cartilage in the body are:

- **Hyaline cartilage**

- **Elastic cartilage**

- **Fibrocartilage**

Fibrocartilage is the strongest of the three types because of the presence of many **collagen** fibers.

Analogy

Three dimensionally, a piece of **any type of cartilage** is similar to a **block of Swiss cheese** in its structure and general consistency. Though cartilage is much stronger, both are solid and flexible. The **cheese** itself is the **matrix** and the **holes** are the **lacunae**.

Location

Intervertebral discs; pubic symphysis; pads within knee joint

Function

Shock absorber in a joint; resists compression

Study Tips

To identify this tissue either under the microscope or from a photograph, look for the following:

- Has the most collagen fibers of any cartilage.

- Collagen fibers often appear in a wavy pattern.

- Chondrocytes are often seen in rows and/or small clusters.

Key to Illustration

1. Lacuna
2. Chondrocyte
3. Nucleus of a chondrocyte
4. Matrix

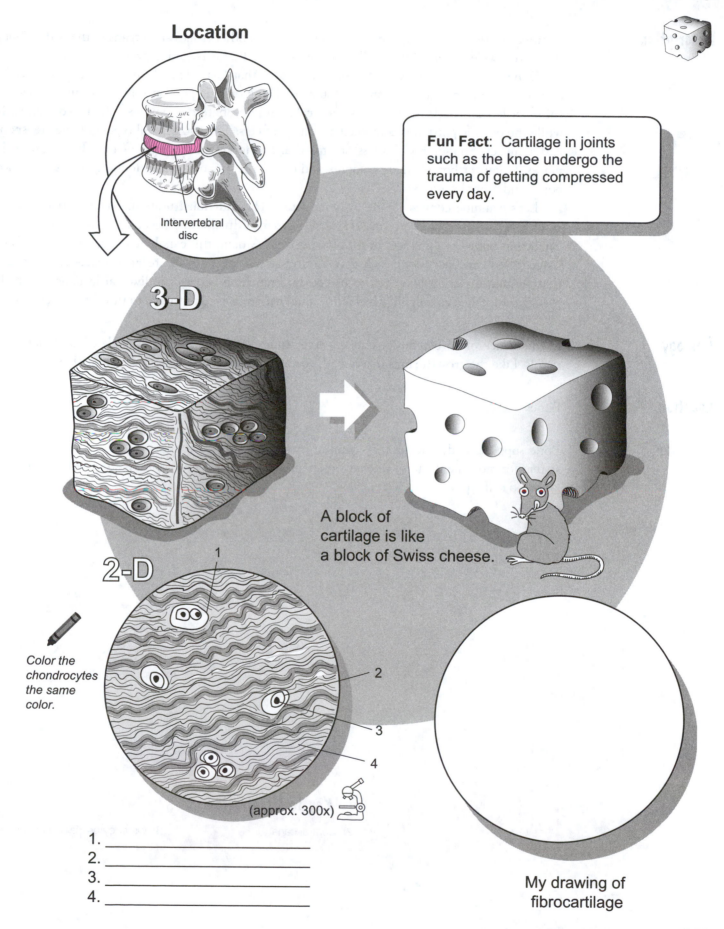

Location

Intervertebral disc

Fun Fact: Cartilage in joints such as the knee undergo the trauma of getting compressed every day.

3-D

A block of cartilage is like a block of Swiss cheese.

2-D

Color the chondrocytes the same color.

(approx. 300x)

1
2
3
4

1. _____
2. _____
3. _____
4. _____

My drawing of fibrocartilage

Bone *(osseous tissue)*

Description

Connective tissues primarily give structural support to other tissues and organs in the body. Though there are a wide variety of types, all are composed of cells, fibers, and matrix.

Bone is a specialized type of connective tissue that has calcified into a hard substance. It is composed of organic and inorganic substances. The inorganic portion that constitutes about two-thirds of bone mass is made of modified calcium phosphate compounds called **hydroxyapatite**, while the organic portion is composed of **collagen** fibers. The two general types of bone are: spongy and compact. Spongy bone is less organized and is found in the ends of long bones and other places. Compact bone is more complex and orderly in structure and is found in the shaft of long bones and other locations.

Let's examine compact bone in more detail. The individual units in compact bone are tall, cylindrical towers called **osteons** (*Haversian systems*). In the middle of each osteon is a **central canal** that serves as a passageway for blood vessels. Around this canal are concentric rings of bony tissue called **lamellae**. Along each of these rings at regular intervals are small spaces called **lacunae** that contain a mature bone cell or **osteocyte**. Branching between individual lacunae are smaller passageways called **canaliculi**, which allow fluid with dissolved nutrients to travel to osteocytes.

Analogy

Each **surface of an osteon** looks like a **tree stump**. Both structures are made of hard, dense materials. Like the **growth rings** in a tree, the osteon has concentric rings called **lamellae**.

Location

Bones

Function

Bone supports body and protects vital organs; provides attachments for muscle to form a lever system for movement; stores calcium compounds and fat. Marrow contains stem cells that produce all blood cell types.

Key to Illustration

1. Osteon
2. Central canal
3. Osteocytes inside lacunae

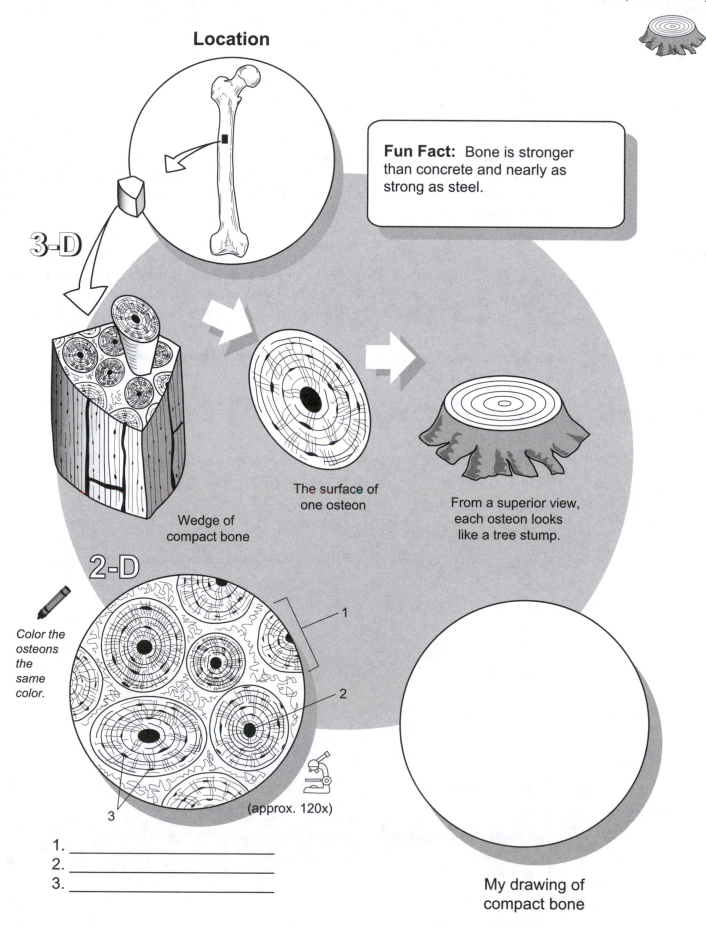

Location

Fun Fact: Bone is stronger than concrete and nearly as strong as steel.

3-D

Wedge of compact bone

The surface of one osteon

From a superior view, each osteon looks like a tree stump.

2-D

Color the osteons the same color.

1

2

3

(approx. 120x)

1. _____

2. _____

3. _____

My drawing of compact bone

Description

There are three different types of muscle tissue:

- **Skeletal muscle**

- **Cardiac muscle**

- **Smooth muscle**

    Skeletal muscle is under conscious control, so it is also referred to as *voluntary* muscle. Each skeletal muscle cell is a long cylinder with a banding pattern, and each band is called a **striation**. Most body cells have only one nucleus per cell, but skeletal muscle has multiple nuclei in each cell—a unique feature.

Analogy

Under the microscope, **skeletal muscle** appears as a bunch of **stacked logs**. **Each log** is equivalent to one **skeletal muscle cell**. Consider them to be birch logs that have a striped pattern on them. These **stripes** are the **striations**. Note that the log doesn't show us the entire tree, just as the image under the microscope doesn't show us the entire cell. This is because the cells are very long.

Location

All the major muscles of the body are composed of skeletal muscle. Examples of skeletal muscle are the biceps brachii, gluteus maximus, and pectoralis major.

Function

Contraction of muscles (*conscious control*)

Study Tips

Under the microscope at higher magnifications, you can use the following landmarks to distinguish skeletal muscle tissue:

- Striations

- Multiple nuclei per cell

- Long cells. (Each cell is so long that you cannot see the ends of it under high magnification.)

Key to Illustration

1. Nuclei within one skeletal muscle cell
2. One skeletal muscle cell
3. Striation

Location

Fun Fact: The human body is composed of more than 600 muscles. Together, they constitute about 40% of body mass.

3-D

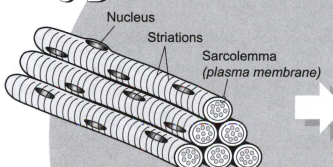

Nucleus

Striations

Sarcolemma
(plasma membrane)

Skeletal muscle cell

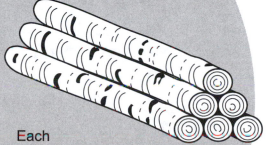

Each
skeletal muscle cell
(or fiber) is like a birch tree log
because it is a long cylinder
with a striped pattern.

2-D

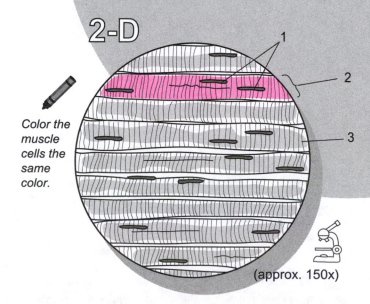

Color the muscle cells the same color.

1

2

3

(approx. 150x)

1. _____

2. _____

3. _____

My drawing of
skeletal muscle

Description

There are three different types of muscle tissue:

- **Skeletal muscle**
- **Cardiac muscle**
- **Smooth muscle**

    Cardiac muscle is under our unconscious control. Each cell has a somewhat cylindrical shape and a single nucleus per cell. One cell connects with another to form a union called an **intercalated disc**. This structure can be seen with a compound microscope. Vertical bands run up and down each cell to form a striped pattern, in which each stripe is a **striation**.

Analogy

The **intercalated discs** are like **two pieces of a jigsaw puzzle** fitting together.

Location

Cardiac muscle is found *only* in the heart.

Function

Contraction of muscles (unconscious control)

Distinguishing Features

Under the microscope at higher magnifications, you can use the following landmarks to distinguish cardiac muscle tissue:

- Striations
- Nucleus appears oval or rounded
- Intercalated discs
- Forking or branching pattern

Key to Illustration

1. Nucleus of one cardiac muscle cell
2. Striation
3. Intercalated disc
4. Individual cardiac muscle cell

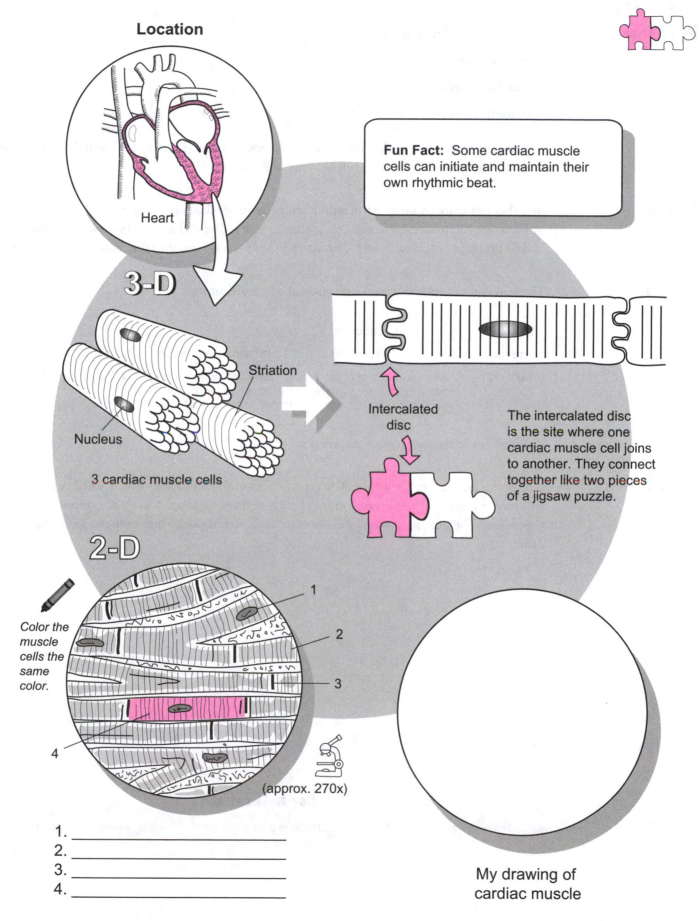

Location

Heart

Fun Fact: Some cardiac muscle cells can initiate and maintain their own rhythmic beat.

3-D

Striation

Nucleus

3 cardiac muscle cells

Intercalated disc

The intercalated disc is the site where one cardiac muscle cell joins to another. They connect together like two pieces of a jigsaw puzzle.

2-D

Color the muscle cells the same color.

1
2
3
4

(approx. 270x)

1. _____
2. _____
3. _____
4. _____

My drawing of cardiac muscle

Description

There are three different types of muscle tissue:

- **Skeletal muscle**
- **Cardiac muscle**
- **Smooth muscle**

Smooth muscle is under unconscious control. It lacks the striations found in the other two types of muscle tissue, and each cell has only one nucleus.

Analogy

In a **sheet of smooth muscle** the **individual cells** are stacked one on top of the other and staggered in their appearance. Each short cell is shaped like elongated ravioli because it is thicker in the middle and tapered on each end. This staggered pattern is similar to the pattern of **bricks in a wall**.

Location

The two types of smooth muscle are **visceral** and **multi-unit:**

Type	Location
Visceral	Walls of hollow organs (stomach, intestines, urinary bladder, etc.)
Multi-unit	Walls of large arteries, trachea, muscles in the iris and ciliary body of the eye, arrector pili muscles that attach to hair follicles

Function

Contraction of muscles (unconscious control)

Study Tips

Under the microscope, it will be difficult to see the cell membrane of individual cells. Instead, you will have to rely on the overall pattern of staggered cells stacked on top of one another to identify this tissue. The general pattern will show up in how the nuclei are arranged with respect to each other.

Key to Illustration

1. Individual smooth muscle cell
2. Nucleus of one smooth muscle cell
3. Plasma membrane

Location

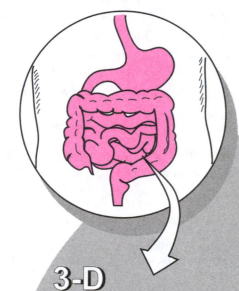

Fun Fact: The drug pitocin is a hormone used to induce labor. It causes the smooth muscle in the uterus to contract.

3-D

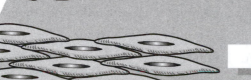

Smooth muscle is often arranged in sheets of cells stacked on top of each other.

The staggered pattern of the cell arrangement looks like bricks in a wall.

2-D

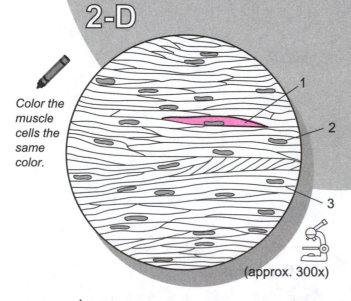

Color the muscle cells the same color.

1
2
3

(approx. 300x)

1. _____
2. _____
3. _____

My drawing of smooth muscle

Description

Neurons, or nerve cells, are one of the fundamental cells in nervous tissue. Of the variety of types of neurons, all share certain features. Surrounding the nucleus of every neuron is a region called the **cell body**. Most of the organelles are found here. Branching out from the cell body are one of two types of processes—**dendrites** or an **axon**. Each neuron has only one axon per cell but may have one or more dendrites. At the end of every axon is the **synaptic knob**, which defines the end of the cell. Surrounding the neurons are various smaller cells that offer structural support and protection that constitute the **neuroglia**.

Analogy

Under the microscope, a **neuron** resembles an **octopus**. The **head** of the octopus is the **cell body** and the **tentacles** are the **cellular processes** (dendrites or axon).

Location

Brain, spinal cord, and peripheral nerves

Function

Conduct nervous (electrical) impulses to other neurons, muscles, or glands to regulate their function

Distinguishing Features

Under the microscope at higher magnifications, you can use the following landmarks to distinguish nervous tissue:

- Cellular processes (unique to nervous tissue)
- Cell body
- Rounded nucleus
- Nucleolus (may or may not be visible depending on slide quality and magnification)

Key to Illustration

1. Dendrites	3. Nucleus	5. Axon
2. Cell body	4. Nucleolus	6. Neuroglia

Location

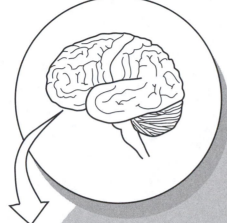

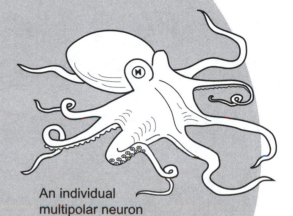

Fun Fact: Neurons are the longest cells in the body. Some are more than 3 feet in length.

3-D

Dendrites

Cell body

Nucleus

Axon

Multipolar neuron

An individual multipolar neuron looks like an octopus.

2-D

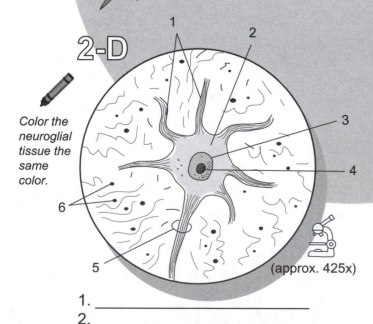

Color the neuroglial tissue the same color.

(approx. 425x)

1. _____
2. _____
3. _____
4. _____
5. _____
6. _____

My drawing of nervous tissue

Notes

Integumentary System

Description

The skin, or **integumentary system**, is the largest organ in the body by total surface area. In the average adult it measures about 21 sq. ft. It is divided into three regional areas: **epidermis**, **dermis**, and **hypodermis** (*subcutaneous region*). The epidermis is the outermost layer and is composed of layers of cells. The dermis lies directly beneath and is subdivided into the upper **papillary region** and the lower **reticular region**. The papillary region consists of loose connective tissue, and the reticular region consists of dense irregular connective tissue. The dermis houses many of the glands and hair follicles within the skin. The hypodermis is below the dermis and is composed of both areolar and adipose connective tissues. Though not a true part of the skin, it is the foundation on which the skin rests. The upper region contains arteries and veins. When a *subcutaneous injection* is given with a *hypodermic* needle, it penetrates into these vessels so the drug can be delivered directly into the bloodstream.

Accessory structures include hair follicles, hair shafts, glands, and sensory receptors. **Hair follicles** produce **hair shafts**. An **arrector pili** muscle is a small cluster of smooth muscle cells connected to hair follicles. When they contract, they pull the hair follicle up, causing "goose bumps." Typical glands are **sebaceous glands** and **sweat glands**. Sebaceous (oil) glands are often connected to hair follicles. They produce an oily substance that lubricates hair shafts.

There are two types of sweat glands: **merocrine** and **apocrine**. Merocrine glands are more common and widely distributed throughout the skin. They secrete a watery secretion called sweat into a duct and release it directly onto the skin surface. This watery film absorbs heat, then evaporates, resulting in cooling the body. Apocrine glands are less common and are connected to hair follicles. They are located in the armpits, around nipples, and in the groin. Their secretion contains both lipids and proteins. When bacteria act on this secretion, the result is body odor.

Sensory receptors include **tactile corpuscles**, **free nerve endings**, and **lamellated corpuscles**. Tactile corpuscles detect mostly light touch, free nerve endings detect mostly pain and temperature changes, and lamellated corpuscles respond to changes in deep pressure.

Study Tips

Layers of Epidermis

- To remember the layers of the epidermis from the outermost to innermost layer, use the following mnemonic: "*Can Lucy Give Some Blood?*"

  (**Note:** This works only with thick skin samples such as in the sole of the foot because the *stratum lucidum* is not present in thin skin.)

- You can remember that the stratum **b**asale is at the **b**ottom of the epidermis because *basale* and *basement* both begin with the letter "**b**."

Function

Physical protection from environment, regulation of body temperature, synthesis of vitamin D, excretion of waste products, prevention of water loss, prevention of invasion by pathogens

Key to Illustration

Layers of Epidermis (Thick Skin)
1. Stratum corneum
2. Stratum lucidum
3. Stratum granulosum
4. Stratum spinosum
5. Stratum basale

Regional Areas of the Skin
6. Epidermis
7. Dermis
8. Papillary region
9. Reticular region
10. Hypodermis (*subcutaneous region*)

Accessory Structures
11. Dermal papillae
12. Tactile corpuscle
13. Free nerve ending
14. Sebaceous (*oil*) gland
15. Hair shaft
16. Hair follicle
17. Lamellated corpuscle
18. Arrector pili muscle
19. Sweat gland duct
20. Merocrine sweat gland
21. Apocrine sweat gland

Layers of Epidermis

Epidermis

1. _____
2. _____
3. _____
4. _____
5. _____

Regional Areas of the Skin

6. _____
7. _____
8. _____
9. _____
10. _____

Sebaceous gland

2-D

3-D

Adipocyte

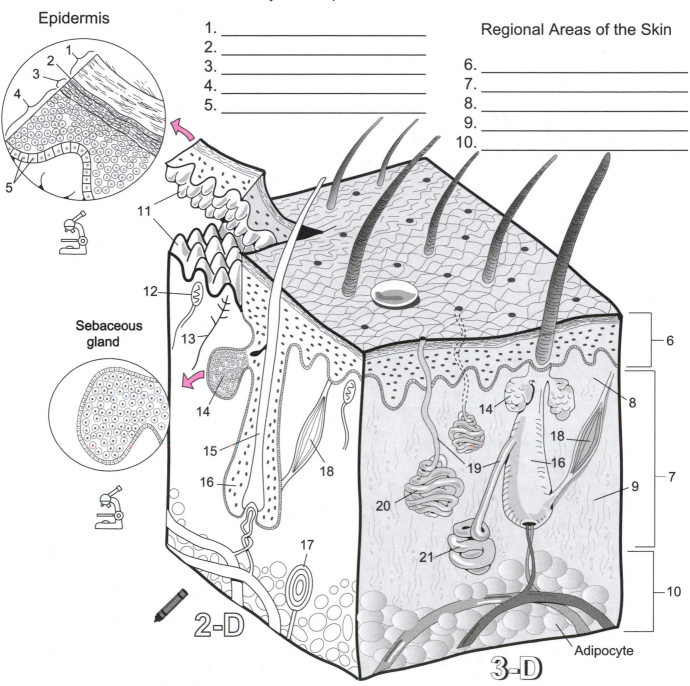

Accessory Structures

11. _____
12. _____
13. _____
14. _____
15. _____
16. _____

17. _____
18. _____
19. _____
20. _____
21. _____

Notes

Skeletal
System

Description

The shaft of the long bone is called the **diaphysis**, while each end is called an **epiphysis**. Covering the diaphysis is a sheath of fibrous connective tissue called **periosteum**, which aids in the attachment of muscles to bone. Covering each epiphysis is a smooth layer of hyaline cartilage that is more generally referred to as **articular cartilage** because it is used to form joints. This smooth surface helps to reduce friction within the joint. Inside the diaphysis is a hollow chamber called the **medullary cavity**. It contains **yellow marrow**, which consists mostly of fatty tissue that acts as a reserve fuel supply for the body. Lining the inside of the medullary cavity is a thin cellular layer called the **endosteum**, which contains both **osteoblasts** and **osteoclasts**.

There are two types of bone within the body—**spongy** and **compact**. Spongy bone can be found within the epiphyses and lining the medullary cavity. Like a sponge, it is more porous, less organized, and contains many open spaces within it. By contrast, compact bone is much more organized and dense. It is much stronger than spongy bone and it constitutes the wall of the diaphysis. In adults, a thin layer of compact bone is also found at the **epiphyseal line**. This marks where the epiphyseal growth plate was located before it ossified.

Study Tip

● Epiphysis is the end of a long bone—"Epiphysis" and "End" both begin with the letter "**e**."

Key to Illustration

1. Proximal epiphysis	5. Spongy bone	9. Endosteum
2. Diaphysis	6. Epiphyseal line	10. Yellow marrow
3. Distal epiphysis	7. Medullary cavity	11. Compact bone
4. Articular cartilage	8. Periosteum	12. Wedge of compact bone

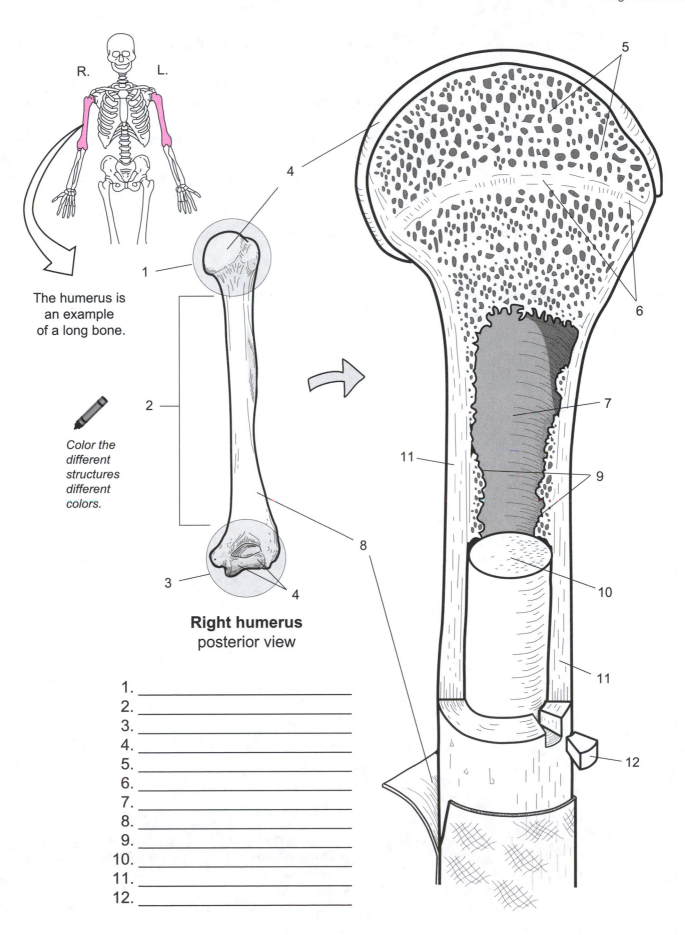

R.   L.

The humerus is
an example
of a long bone.

Color the
different
structures
different
colors.

Right humerus
posterior view

1. _____
2. _____
3. _____
4. _____
5. _____
6. _____
7. _____
8. _____
9. _____
10. _____
11. _____
12. _____

Description

There are two types of bone within the body—**spongy** and **compact**. Spongy bone can be found within the epiphyses and lining the medullary cavity. Like a sponge, it is more porous, less organized, and contains many open spaces within it. By contrast, compact bone is much more organized and dense. It is much stronger than spongy bone, and it constitutes the wall of the diaphysis.

Let's examine compact bone in more detail. The individual units in compact bone are tall, cylindrical towers called **osteons** (*Haversian systems*). In the middle of each osteon is a **central canal** that serves as a passageway for blood vessels. Around this canal are concentric rings of bony tissue called **lamellae**. Along each of these rings at regular intervals are small spaces called **lacunae**, which contain a mature bone cell or **osteocyte**. Branching between individual lacunae are smaller passageways called **canaliculi**, which allow fluid with dissolved nutrients to travel to osteocytes.

Analogy

Each **surface of an osteon** looks like a **tree stump**. Both structures are made of hard, dense materials. Like the **growth rings** in a tree, the osteon has concentric rings called **lamellae**.

Key to Illustration

1. Osteon (*Haversian system*)
2. Lamellae
3. Collagen fibers
4. Periosteum
5. Spongy bone
6. Communicating canal
7. Central canal
8. Lacuna
9. Osteocyte
10. Canaliculi

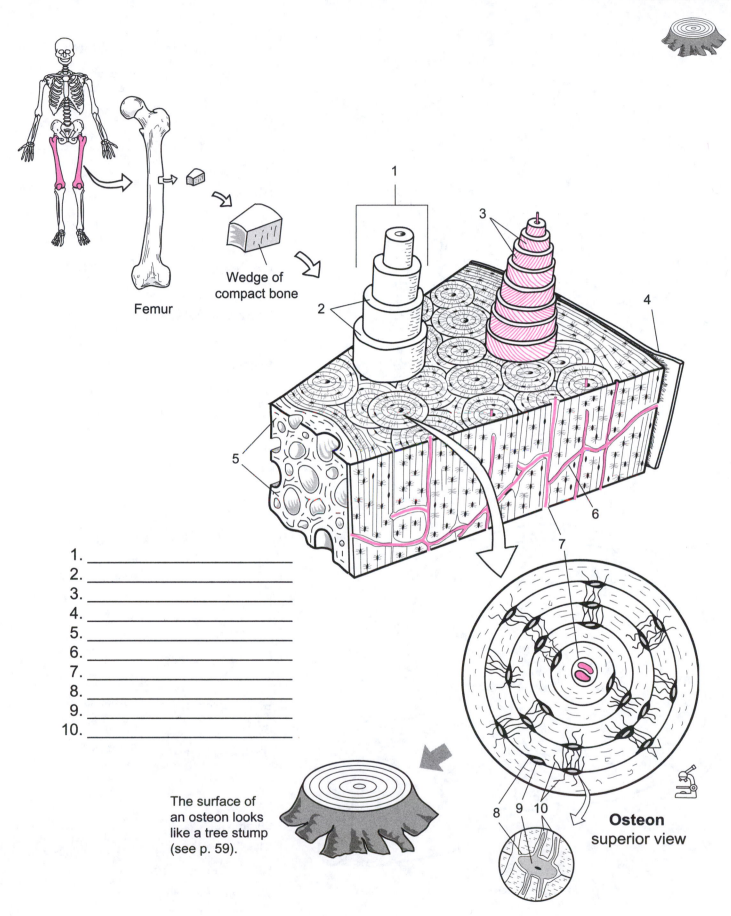

Femur

Wedge of
compact bone

1. _____
2. _____
3. _____
4. _____
5. _____
6. _____
7. _____
8. _____
9. _____
10. _____

The surface of
an osteon looks
like a tree stump
(see p. 59).

Osteon
superior view

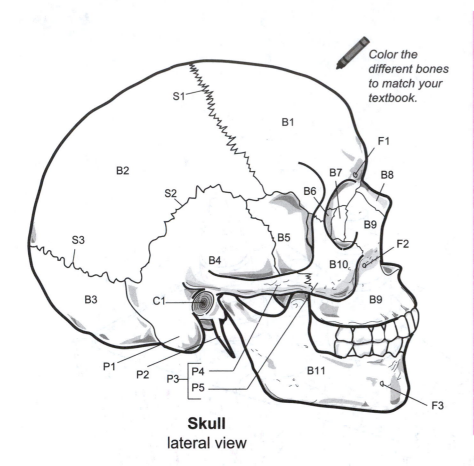

Color the different bones to match your textbook.

Skull
lateral view

Key to Lateral View

Bones (B)

B1 Frontal bone

B2 Parietal bone

B3 Occipital bone

B4 Temporal bone

B5 Sphenoid

B6 Ethmoid

B7 Lacrimal bone

B8 Nasal bone

B9 Maxilla

B10 Zygomatic bone

B11 Mandible

Sutures (S)

S1 Coronal suture

S2 Squamous suture

S3 Lambdoid suture

Processes, Projections (P)

P1 Mastoid process

P2 Styloid process

P3 Zygomatic arch

P4 Zygomatic process of temporal bone

P5 Temporal process of zygomatic bone

Foramina (F)

F1 Supraorbital foramen

F2 Infraorbital foramen

F3 Mental foramen

Canal (C)

C1 External acoustic meatus (*canal*)

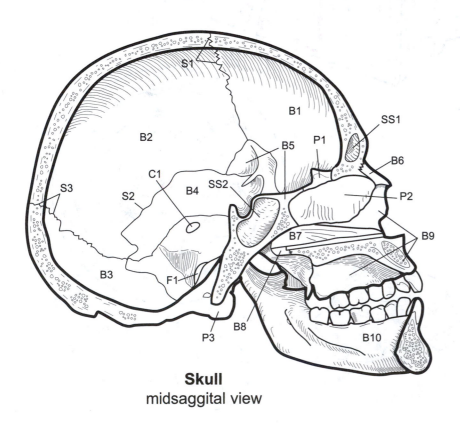

Skull
midsaggital view

Key to Midsaggital View

Bones (B)

B1 Frontal bone

B2 Parietal bone

B3 Occipital bone

B4 Temporal bone

B5 Sphenoid

B6 Nasal bone

B7 Vomer

B8 Palatine bone

B9 Maxilla

B10 Mandible

Sutures (S)

S1 Coronal suture

S2 Squamous suture

S3 Lambdoid suture

Processes, Projections (P)

P1 Crista galli (*of ethmoid*)

P2 Perpendicular plate (*of ethmoid*)

P3 Occipital condyle

Foramina (F)

F1 Jugular foramen

Sinuses (SS)

SS1 Frontal sinus

SS2 Sphenoidal sinus

Canal (C)

C1 Internal acoustic meatus (*canal*)

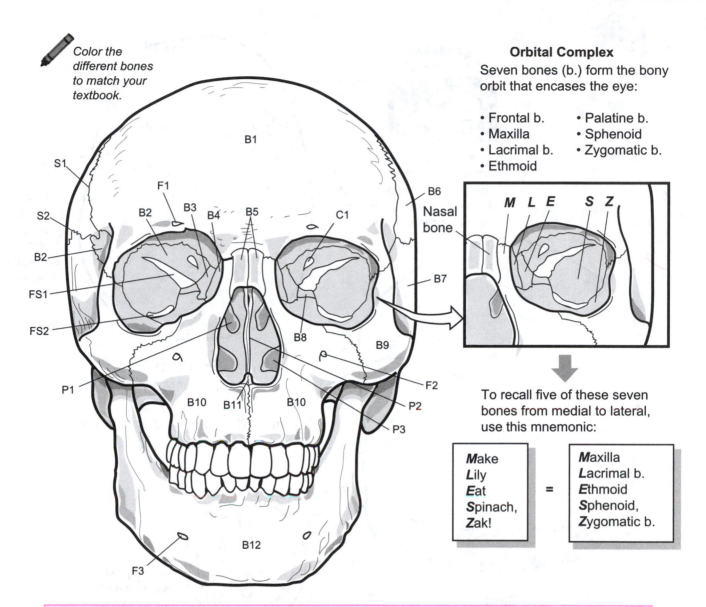

Color the different bones to match your textbook.

Orbital Complex

Seven bones (b.) form the bony orbit that encases the eye:

- Frontal b.
- Maxilla
- Lacrimal b.
- Ethmoid
- Palatine b.
- Sphenoid
- Zygomatic b.

To recall five of these seven bones from medial to lateral, use this mnemonic:

Make **L**ily **E**at **S**pinach, **Z**ak!	=	**M**axilla **L**acrimal b. **E**thmoid **S**phenoid, **Z**ygomatic b.

Key to Illustration

Bones (B)

B1 Frontal bone

B2 Sphenoid

B3 Ethmoid

B4 Lacrimal bone

B5 Nasal bone

B6 Parietal bone

B7 Temporal bone

B8 Palatine bone

B9 Zygomatic bone

B10 Maxilla

B11 Vomer

B12 Mandible

Sutures (S)

S1 Coronal suture

S2 Squamous suture

Foramina (F)

F1 Supraorbital foramen

F2 Infraorbital foramen

F3 Mental foramen

Canal (C)

C1 Optic canal

Processes, Projections (P)

P1 Middle nasal concha

P2 Perpendicular process of ethmoid

P3 Inferior nasal concha

Fissures (FS)

FS1 Superior orbital fissure

FS2 Inferior orbital fissure

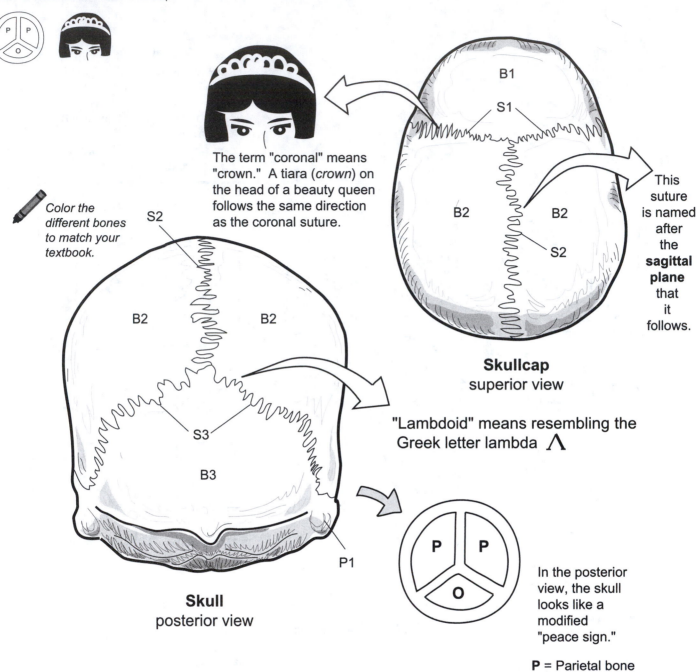

The term "coronal" means "crown." A tiara (*crown*) on the head of a beauty queen follows the same direction as the coronal suture.

Color the different bones to match your textbook.

This suture is named after the **sagittal plane** that it follows.

Skullcap
superior view

"Lambdoid" means resembling the Greek letter lambda Λ

Skull
posterior view

In the posterior view, the skull looks like a modified "peace sign."

P = Parietal bone
O = Occipital bone

Key to Illustration

Bones (B)	Sutures (S)	Processes, Projections (P)
B1 Frontal bone	S1 Coronal suture	P1 Mastoid process
B2 Parietal bone	S2 Sagittal suture	
B3 Occipital bone	S3 Lambdoid suture	

Simplify what you see, use the **sella turcica** and foramina as landmarks.

Sella turcica

F1
F2
F3
F4
F5

Sella turcica (saddle-like structure)

Foramen magnum (largest hole)

From the lateral view, the **sella turcica** resembles a horse's saddle.

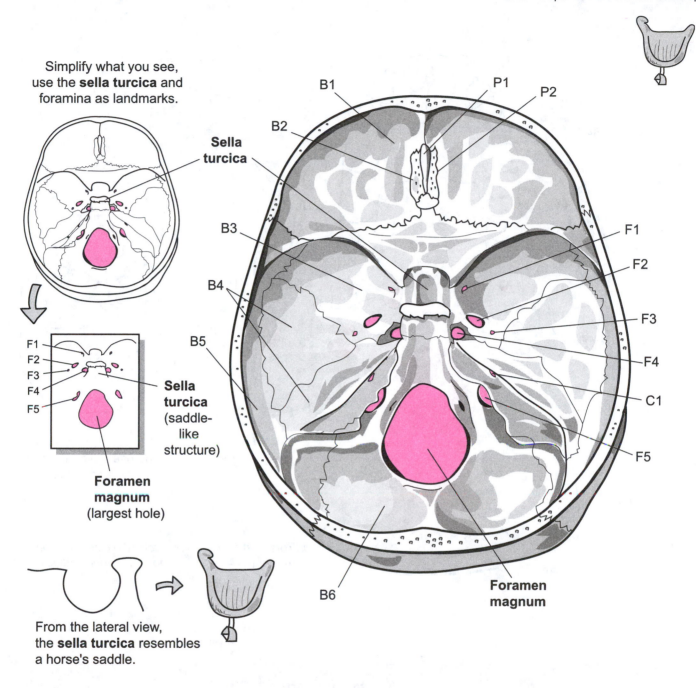

B1
B2
P1
P2
B3
F1
F2
B4
F3
B5
F4
C1
F5
B6
Foramen magnum

Key to Illustration

Bones (B)

B1 Frontal bone

B2 Ethmoid

B3 Sphenoid

B4 Temporal bone

B5 Parietal bone

B6 Occipital bone

Processes, Projections (P)

P1 Crista galli

P2 Cribriform plate

Canal (C)

C1 Internal acoustic meatus

Foramina (F)

F1 Foramen rotundum

F2 Foramen ovale

F3 Foramen spinosum

F4 Foramen lacerum

F5 Jugular foramen

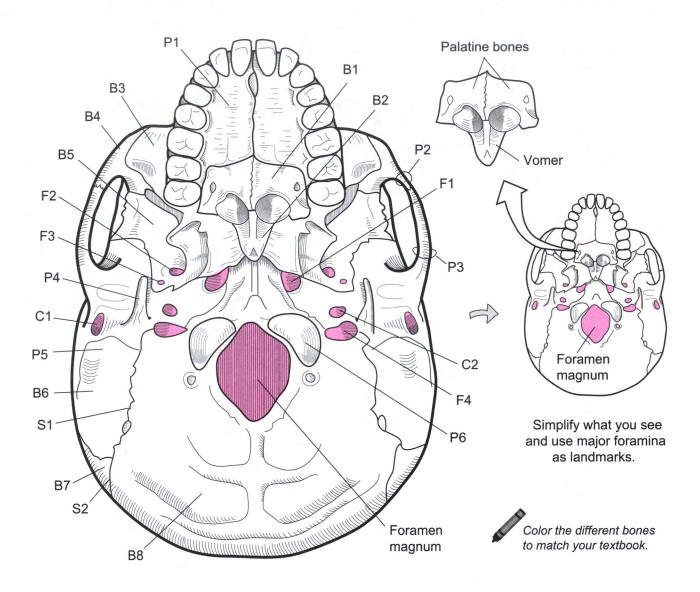

Palatine bones

Vomer

Foramen magnum

Simplify what you see and use major foramina as landmarks.

Color the different bones to match your textbook.

Foramen magnum

Key to Illustration

Bones (B)

B1 Palatine bone

B2 Vomer

B3 Maxilla

B4 Zygomatic bone

B5 Sphenoid

B6 Temporal bone

B7 Parietal bone

B8 Occipital bone

Sutures (S)

S1 Occipitomastoid suture

S2 Lambdoid suture

Foramina (F)

F1 Foramen lacerum

F2 Foramen ovale

F3 Foramen spinosum

F4 Jugular foramen

Processes, Projections (P)

P1 Palatine process of maxilla

P2 Temporal process of zygomatic bone

P3 Zygomatic process of temporal bone

P4 Styloid process (*temporal bone*)

P5 Mastoid process (*temporal bone*)

P6 Occipital condyle

Canal (C)

C1 External acoustic meatus (*canal*)

C2 Carotid canal

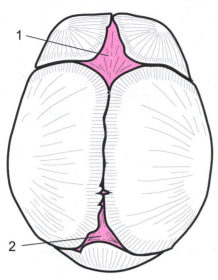

Fetal skull
superior view

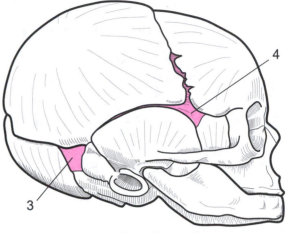

Color the different bones to match your textbook.

Fetal skull
lateral view

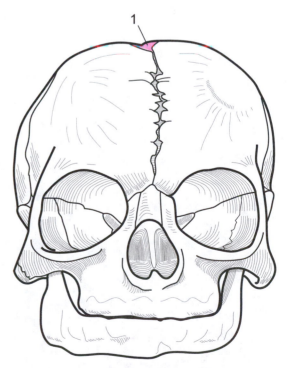

Fetal skull
frontal view

Fontanels are flat areas of fibrous connective tissue between bones of the skull that have not completed the ossification process. There are 4 major fontanels in the fetal skull. What is referred to as the "soft spot" on an infant's skull is actually the **anterior fontanel**, which typically closes by 2 years of age.

Fontanels

1. _____
2. _____
3. _____
4. _____

Key to Illustration

1. Anterior fontanel

2. Posterior fontanel

3. Mastoid fontanel

4. Sphenoid fontanel

Description

The temporal bone is located on the side of the skull. It articulates with the sphenoid, parietal, and occipital bones. The zygomatic process of the temporal bone articulates with the temporal process of the zygomatic bone to form the zygomatic arch.

Analogy

The **temporal bone in the lateral view** is like the **head of a rooster**. The **squamous portion** (*squama*) is the **rooster's comb**, the **external acoustic meatus** is the **rooster's eye**, the **styloid process** is the **rooster's beak**, and the **mastoid process** is the **rooster's wattles**.

Study Tips

Palpate (*touch*): Feel behind your ear to locate the large bump—the **mastoid process**.

- The **styloid process** is often broken off of a real skull because it is a delicate structure. Do not be surprised if you cannot locate it on a *real* skull, but a good-quality plastic skull will have this structure.

- Squamous means "scale-like," which accurately describes the flat, *squamous* portion of the temporal bone.

- Processes and other structures are sometimes named after the bones with which they articulate: *e.g.*, zygomatic process of temporal bone and temporal process of zygomatic bone.

Key to Illustration

1. Squamous part *(squama)*
2. Zygomatic process
3. External acoustic meatus
4. Mastoid process
5. Styloid process

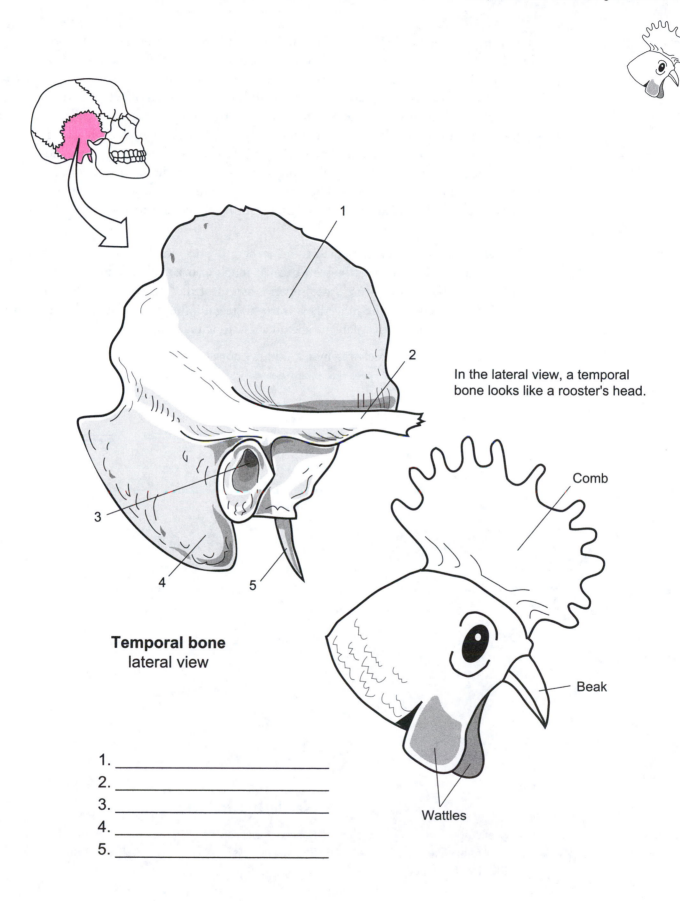

In the lateral view, a temporal bone looks like a rooster's head.

Temporal bone
lateral view

Comb

Beak

Wattles

1. _____

2. _____

3. _____

4. _____

5. _____

Description

The **ethmoid** is embedded in the skull and located behind the bridge of the nose. The entire bone is not visible. Only specific portions of it can be seen, which makes it difficult to visualize how it fits into the skull. On its superior aspect are several important structures. A flat plate of bone called the **cribriform plate** has many small holes in it called **olfactory foramina** that allow olfactory nerves to pass from the olfactory organ to the brain. A partition called the **crista galli** separates the cribriform plate into left and right halves. In the anterior view, a long vertical plate of bone called the **perpendicular plate** forms part of the nasal septum. On either side of this plate are tiny, curved, bony projections called the **superior** and **middle nasal concha**. The sides of the bone consist of the lateral masses. These structures contain a network of interconnected, hollow chambers called **ethmoid air cells**. Together, these air cells constitute the **ethmoid sinuses**.

Analogy

- The entire bone can be compared to an iceberg floating in the water. Like the iceberg, the entire ethmoid bone cannot be seen because much of it is embedded in the skull. The tip of the iceberg that can be seen is the superior aspect of the ethmoid bone, which looks like a door hinge folded flat. The hinge itself is the crista galli, the metal plate is the cribriform plate, and the screw holes in the plate are the olfactory foramina.

- The crista galli more closely resembles a shark's dorsal fin.

Location

Skull

Key to Illustration

1. Crista galli	4. Superior nasal concha	6. Middle nasal concha
2. Cribriform plate	5. Ethmoid air cells	7. Perpendicular plate
3. Olfactory foramina		

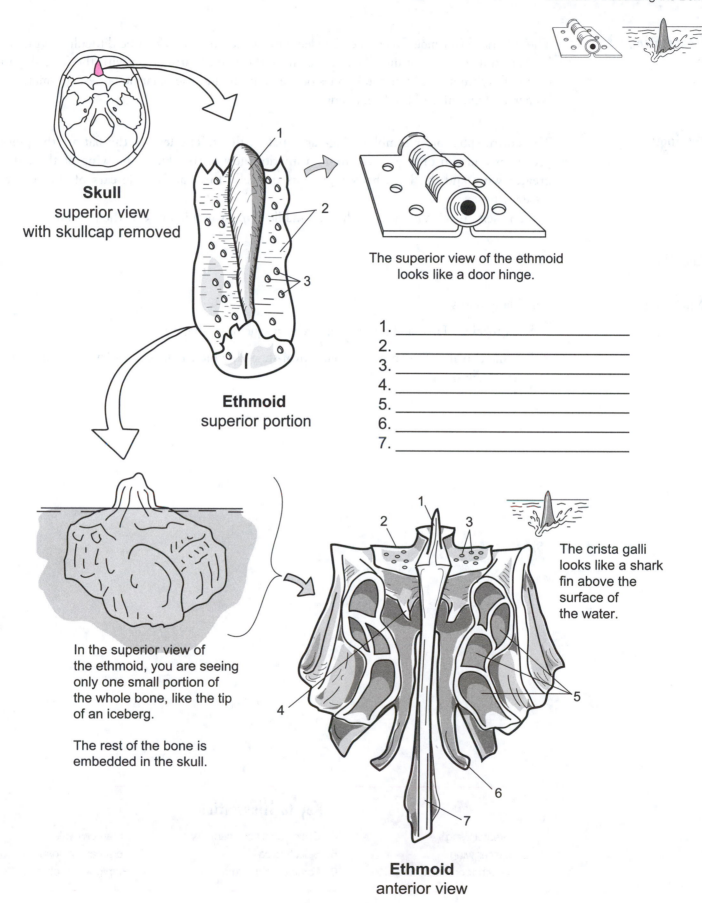

Skull
superior view
with skullcap removed

Ethmoid
superior portion

The superior view of the ethmoid
looks like a door hinge.

1. _____
2. _____
3. _____
4. _____
5. _____
6. _____
7. _____

In the superior view of
the ethmoid, you are seeing
only one small portion of
the whole bone, like the tip
of an iceberg.

The rest of the bone is
embedded in the skull.

The crista galli
looks like a shark
fin above the
surface of
the water.

Ethmoid
anterior view

87

Description

The sphenoid (*sphenoid bone*) is embedded within the skull, so it can be difficult to visualize. To be seen in its entirety, it must be removed from the skull. It articulates with the frontal, parietal, occipital, ethmoid and temporal bones of the cranium and the palatine bones, zygomatic bones, vomer, and maxillae of the facial bones.

Analogy

The **entire sphenoid** resembles a **big-eared bat in flight**. The **legs of the bat** are the **pterygoid processes** and can only be seen from an inferior view of the skull. The **wing of the bat** is the **greater wing** and the **bat's body** represents the **sella turcica**. The big **ears of the bat** are the **lesser wings**.

The **sella turcica** resembles a **horse's saddle**. This structure protects the pituitary gland.

Location

Skull

Study Tips

Good landmarks:

- **Sella turcica.** This saddle-like structure is unique in its shape.

- **Foramen ovale.** This hole is usually oval-shaped, which makes it easier to distinguish from other foramina.

Key to Illustration

1. Greater wing	4. Optic canal (foramen)	7. Foramen ovale
2. Lesser wing	5. Sella turcica	8. Foramen spinosum
3. Superior orbital fissure	6. Foramen rotundum	9. Pterygoid process

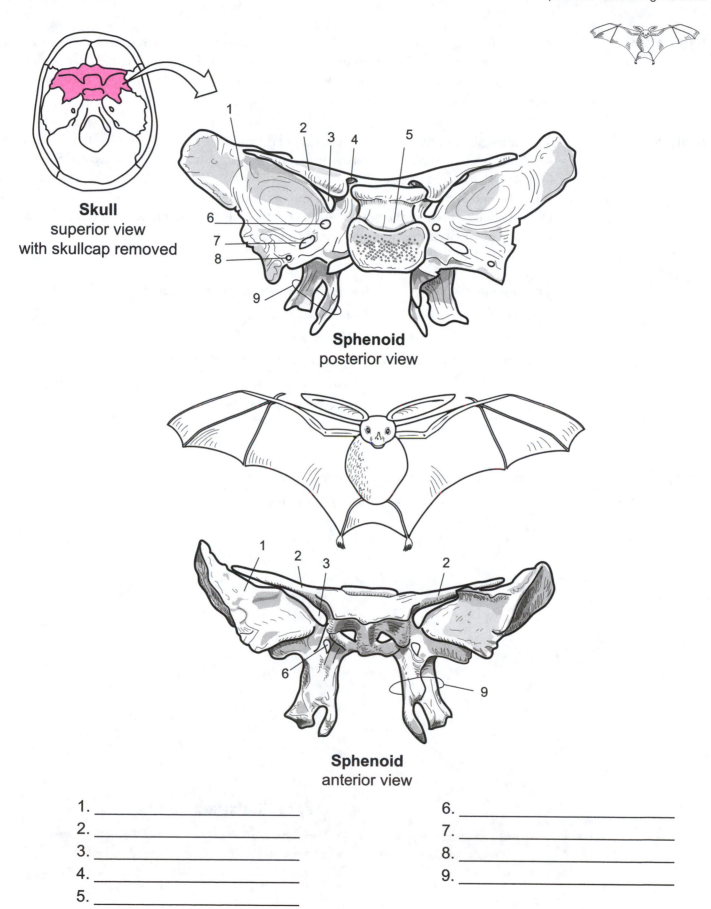

Skull
superior view
with skullcap removed

Sphenoid
posterior view

Sphenoid
anterior view

1. _____

2. _____

3. _____

4. _____

5. _____

6. _____

7. _____

8. _____

9. _____

Description

A **foramen** (sing., *foramina* = plural) is a hole in bone through which a structure such as a blood vessel or nerve passes. The **sphenoid** bone contains many foramina. At first glance they appear like many small, indistinguishable holes. The challenge is to be able to differentiate one from the other.

Analogy

This analogy will link five different sphenoidal foramina. The **sella turcica** looks like a **horse's saddle**. Imagine a miniature cowboy sitting on the sella turcica. He is named **ROS** for the three foramina (F.) on either side of him. From medial to lateral, they are as follows: F. **R**otundum, F. **O**vale, and F. **S**pinosum. His **Legs** go through foramen **Lacerum** and his arms go through the **Optic** foramen. Because the optic nerve passes through this opening, imagine ROS holding onto the optic nerves as if they were the reins of a horse.

Function

Foramen (F.)	Blood Vessel / Nerve Passing Through
Optic F.	Cranial nerve II (optic) and ophthalmic artery
F. rotundum	Maxillary branch of cranial nerve V (*trigeminal*)
F. ovale	Mandibular branch of cranial nerve V (*trigeminal*)
F. spinosum	Blood vessels to membranes around central nervous system
F. lacerum	Branch of ascending pharyngeal artery

Study Tips

- The **foramen ovale** tends to be shaped like an oval.
- The **F. spinosum** is the smallest of the foramina.

Key to Illustration

1. Optic foramen
2. Foramen rotundum
3. Foramen ovale
4. Foramen spinosum
5. Foramen lacerum
6. Sella turcica

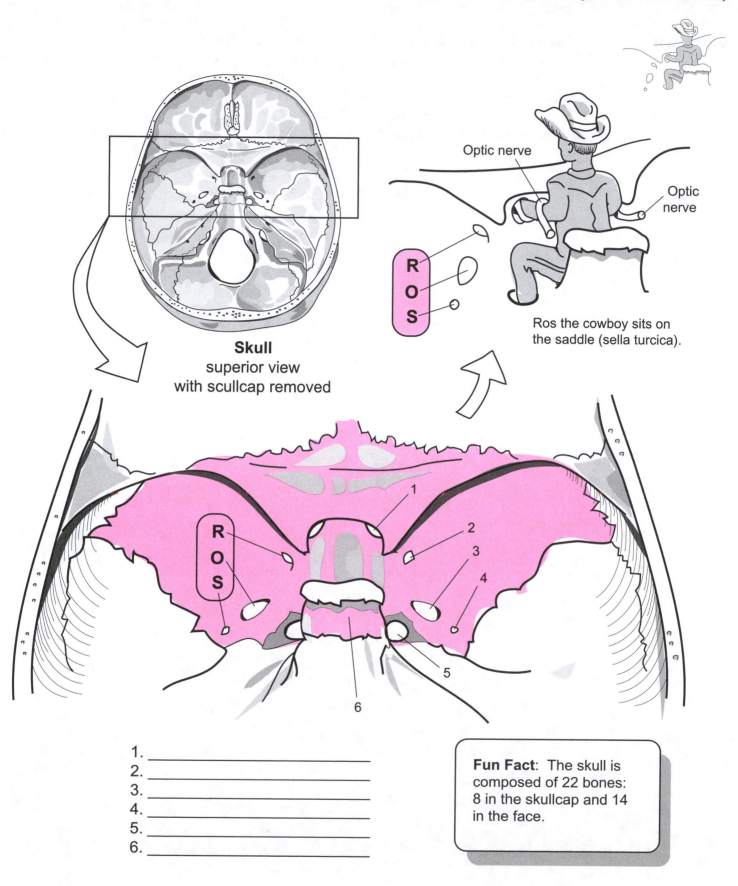

Skull
superior view
with scullcap removed

Optic nerve

Optic nerve

Ros the cowboy sits on
the saddle (sella turcica).

1. _____
2. _____
3. _____
4. _____
5. _____
6. _____

Fun Fact: The skull is
composed of 22 bones:
8 in the skullcap and 14
in the face.

Description

The skull has two, small **palatine bones**. Two major parts of this bone are the **horizontal plate** and the **perpendicular plate**. The horizontal plate articulates with the maxillae to form the posterior portion of the hard palate in the roof of the mouth. When the horizontal plates of both bones touch, they form a narrow ridge called the **nasal crest**, that articulates with the vomer. The perpendicular plate runs vertically and articulates with the maxillae, ethmoid, sphenoid, and inferior nasal concha. The **orbital process** is located on top of the perpendicular plate to mark the most superior part of the bone. It forms a small part of the posterior portion of the orbit.

Analogy

In the anterior view, each **palatine bone** looks like a **letter "L."** Both bones fused together look like **two mirror-image letter L's touching each other**.

Location

Skull

Key to Illustration

1. Horizontal plate
2. Nasal crest
3. Perpendicular process
4. Orbital process

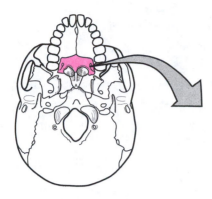

Skull
inferior view with
mandible removed

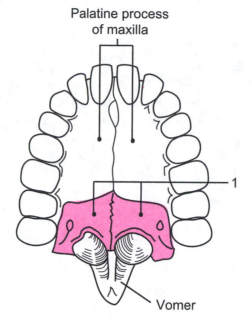

Palatine process
of maxilla

1

Vomer

Palatine bones
inferior view

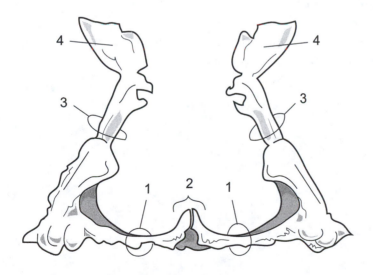

4

4

3

3

1 2 1

Palatine bones
anterior view

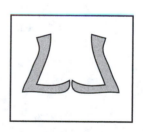

In the anterior view, each palatine
bone resembles a letter "L." The two
bones fused together are like two
mirror-image L's touching each other.

1. _____
2. _____
3. _____
4. _____

Description

The vertebral column contains a total of **24** vertebrae of **three** different types:

- **Cervical** (7)
- **Thoracic** (12)
- **Lumbar** (5)

In total there are **7** cervical, **12** thoracic, and **5** lumbar vertebrae. To remember the total number of each type of vertebrae, think of meal times:

- **Breakfast** at **7:00 a.m.**
- **Lunch** at **12:00 noon**
- **Dinner** at **5:00 p.m.**

To remember the **sacrum** and **coccyx**, think of having 2 snacks in the evening:

- 1st snack = sacrum
- 2nd snack = coccyx

The **sacrum** results from the fusion of five vertebrae. On the top, in the anterior view, is a ridge of bone called the **sacral promontory**. This is an important landmark for a female's pelvic exam. The series of holes running through the bone are called the **sacral foramina**—nerves of the sacral plexus pass through them. On the posterior surface, there is an opening called the **sacral hiatus** that leads into a long passageway called the **sacral canal**. Nerves from the spinal cord run through this canal.

The **coccyx** (*tailbone*) is located inferior to the sacrum and consists of 3–5 bones. It serves as an anchor point for muscles, tendons, and ligaments.

Total Number of Each Type of Vertebrae

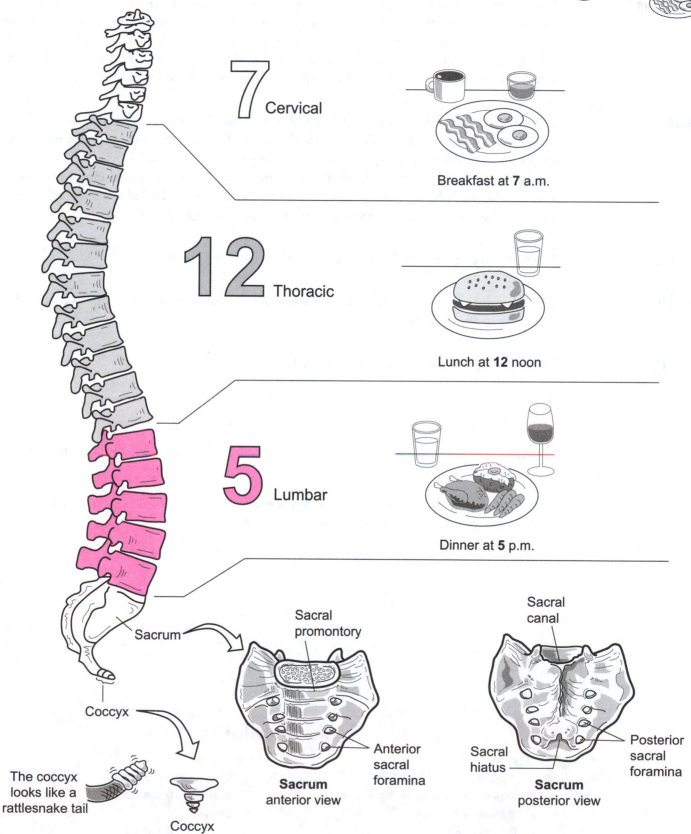

7 Cervical

Breakfast at **7** a.m.

12 Thoracic

Lunch at **12** noon

5 Lumbar

Dinner at **5** p.m.

Sacrum

Coccyx

Sacral promontory

Sacral canal

Anterior sacral foramina

Posterior sacral foramina

Sacral hiatus

Sacrum anterior view

Sacrum posterior view

The coccyx looks like a rattlesnake tail

Coccyx

Description

The vertebral column contains a total of 24 vertebrae of three different types: cervical (7), thoracic (12), and lumbar (5). The first two cervical vertebrae at the top of the vertebral column are referred to as the **atlas** (*cervical 1 or C1*) and the **axis** (*cervical 2 or C2*). Like the Greek god Atlas held up the earth, the atlas vertebra is positioned at the base of the globe-like skull. The atlas is designed to pivot on the axis, which permits you to turn your head from side to side.

Analogy

The **atlas** (C1) resembles a **turtle's head with eyeglasses**. The **anterior arch** is the handle of the eyeglasses, the **superior articular facets** are the lenses of the glasses, the **transverse process** is the arm of the eyeglasses, and the **posterior arch** is the smile on the turtle's face.

The **axis** (C2) resembles a **football player grasping a football**. The **odontoid process** is the football player's helmet, the **superior articular facet** is the football player's shoulder pad, the **lamina** is the **forearm**, and the **spinous process** is the **hands grasping the football**.

Location

The first two cervical vertebrae at the top of the vertebral column

Function

The atlas and axis together form a pivot joint. When you turn your head from side to side, the atlas is rotating on the more stationary axis.

Study Tips

The following are good landmarks for these bones:

Atlas	Axis
• transverse foramen	• transverse foramen
• **large** superior articular facets	• odontoid process (*dens*) (unique to axis only)
	• **large** superior articular facets

Note that *only* cervical vertebrae have a **transverse foramen**. This makes them easy to distinguish from thoracic or lumbar vertebrae.

Key to Illustration

Atlas
1. Anterior arch
2. Superior articular facet
3. Transverse process
4. Posterior arch
5. Transverse foramen

Axis
6. Odontoid process (*dens*)
7. Superior articular facet
8. Lamina
9. Bifid spinous process
10. Transverse foramen

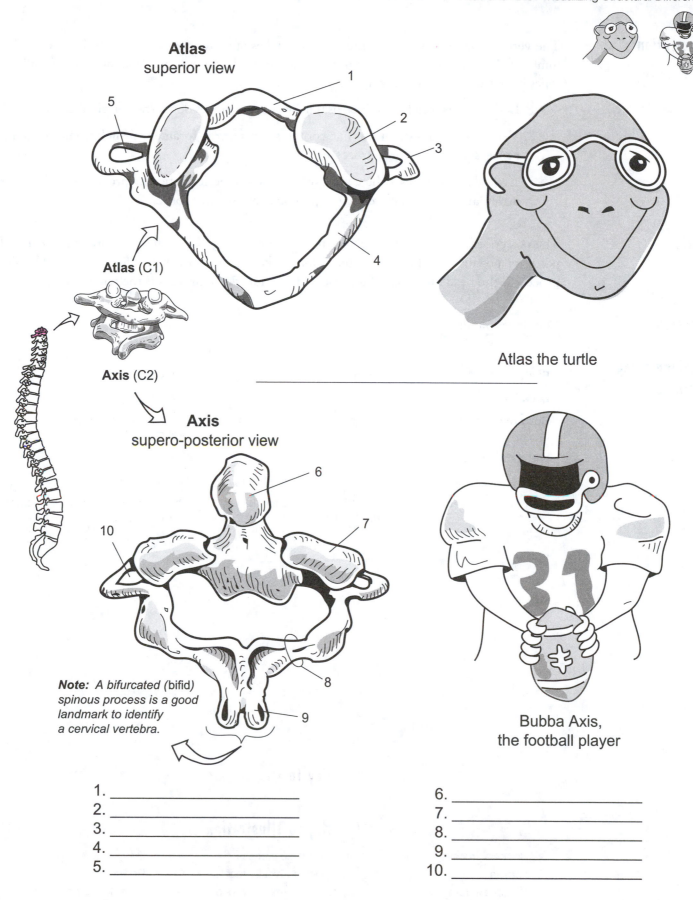

Atlas
superior view

5

1

2

3

4

Atlas (C1)

Axis (C2)

Axis
supero-posterior view

6

10

7

8

9

Note: *A bifurcated (bifid) spinous process is a good landmark to identify a cervical vertebra.*

Atlas the turtle

Bubba Axis,
the football player

1. _____
2. _____
3. _____
4. _____
5. _____

6. _____
7. _____
8. _____
9. _____
10. _____

Description

The vertebral column contains three different types of vertebrae: cervical (7), thoracic (12), and lumbar (5). Each type has its own unique features to distinguish one from another, yet all of the types have three basic features in common:

1. **Body**—bears weight and increases in size as one moves down the vertebral column.

2. **Vertebral arch**—structure that contains a **vertebral foramen**, **pedicles**, **lamina**, **spinous process**, and **transverse processes**.

3. **Articular processes**—there are two of these—**superior articular process** and the **inferior articular process** that are used to join one vertebra to another.

Analogy

The posterior portion of a **thoracic vertebra** looks like a **goose with wings** *arched forward*. The posterior portion of a **lumbar vertebra** looks like a goose with wings *horizontal*. The **spinous process** is the **head and neck of a goose** and the **transverse process** is the **wing of a goose**.

Location

Vertebral column

Distinguishing Features

Feature	Thoracic	Lumbar
Location	Chest	Lower back
Body of vertebra	Medium-sized, heart-shaped; facets for ribs	Largest diameter, thicker, oval-shaped
Vertebral foramen	Medium-sized	Smaller-sized
Spinous process	Long, slender; points inferiorly	Broad; flat; blunt
Transverse process	10 of 12 have facets for rib articulations	Short; narrower, no articular facets or transverse foramina

Key to Illustration

1. Spinous process
2. Lamina
3. Vertebral foramen
4. Transverse process
5. Superior articular facet
6. Superior articular process
7. Body
8. Pedicle
9. Facet for rib articulation

Fun Fact: A human neck has the same number of vertebrae as a giraffe.

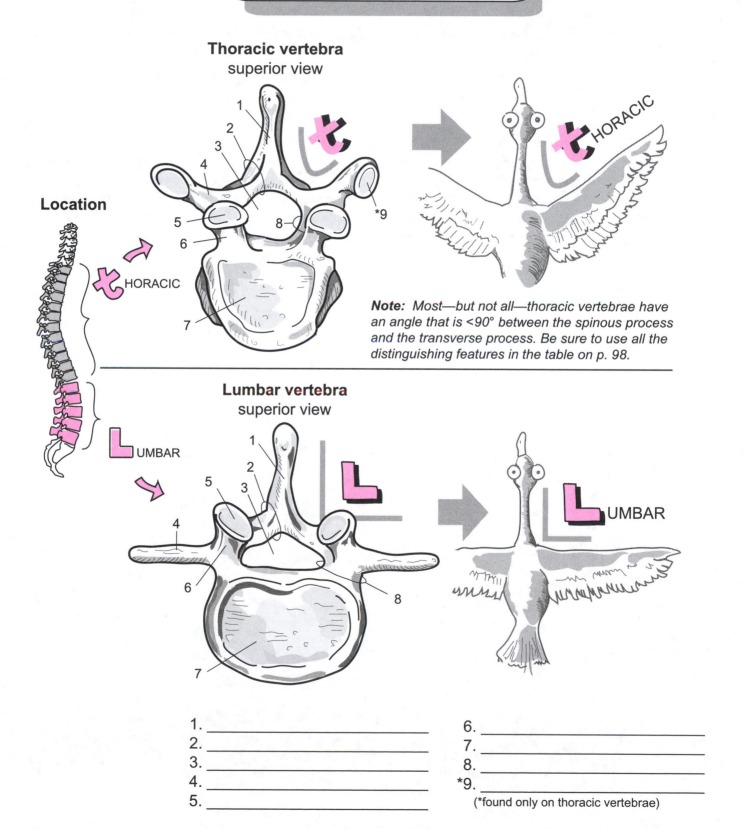

Thoracic vertebra
superior view

Location

HORACIC

Note: *Most—but not all—thoracic vertebrae have an angle that is <90° between the spinous process and the transverse process. Be sure to use all the distinguishing features in the table on p. 98.*

Lumbar vertebra
superior view

LUMBAR

1. _____
2. _____
3. _____
4. _____
5. _____

6. _____
7. _____
8. _____
*9. _____

(*found only on thoracic vertebrae)

Description

The vertebral column contains three different types of vertebrae: cervical (7), thoracic (12), and lumbar (5). Each type has its own unique features to distinguish one from another, yet all of the types have three basic features in common:

1. **Body**—bears weight and increases in size as one moves down the vertebral column.

2. **Vertebral arch**—structure that contains a **vertebral foramen, pedicles, lamina, spinous process,** and **transverse processes.**

3. **Articular processes**—there are two of these—**superior articular process** and the **inferior articular process** that are used to join one vertebra to another.

Analogy

In the postero-lateral view, the **thoracic vertebra** looks like a **giraffe**. The **giraffe's snout** is the **spinous process**, the **giraffe's ears** are the **transverse processes**, and the **giraffe's horns** are the **superior articulating processes**. The **giraffe's cheek** is the **inferior articulating process**.

In the lateral view, a **lumbar vertebra** looks like the **head of a moose**. The **moose's snout** is the **spinous process**, the **moose's horns** are the **superior articulating processes**, and the **moose's bell** is the **inferior articulating process**.

Location

Vertebral column

Key to Illustration

1. Body
2. Superior articular process
3. Superior articular facet
4. Transverse process
5. Spinous process
6. Inferior articular process
7. Inferior articular facet

Location

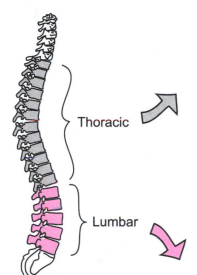

Thoracic

Lumbar

"Thoracic giraffe"
In this postero-lateral view, the posterior portion of a lumbar vertebra looks like the head of a giraffe.

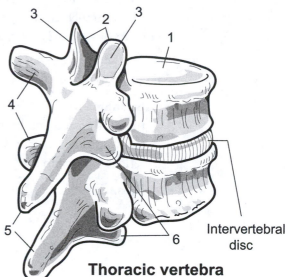

Intervertebral disc

Thoracic vertebra
postero-lateral view

1. _____
2. _____
3. _____
4. _____
5. _____
6. _____
7. _____

Bell

"Lumbering moose"
In the lateral view, the posterior portion of a lumbar vertebra looks like the head of a moose.

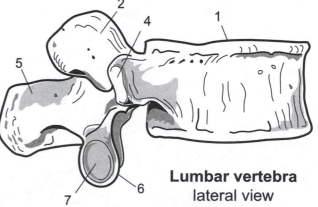

Lumbar vertebra
lateral view

Description

The **sternum** is commonly called the breastbone. It is divided into three parts: manubrium, body, and xiphoid process. The manubrium articulates with the clavicles and the costal cartilages of the first pair of ribs. The body is the largest part and articulates with the ribs through costal cartilages. The xiphoid process is the smallest part and does not articulate with another bone.

There are 12 pairs of **ribs**. The first seven are called **true ribs** because they directly attach to the sternum through costal cartilages. The **false ribs** (8–12) do not directly attach to the sternum. The last two ribs (11–12) are called "floating" because they have no connection to the sternum.

Study Tips

Palpate (*feel by touch*):

- **Sternum:** You can easily feel the middle of the **manubrium** and **body** of your sternum. The small tip at the end marks the **xiphoid process**. During CPR training, students are instructed to avoid doing compressions on the xiphoid process because it can break off easily and cause serious damage to the liver.

- **Ribs:** The body of the ribs can be felt on the lateral surface of the thoracic cage.

Key to Illustration

Rib (R)	Sternum (S)
R1. Head	S1. Jugular notch
R2. Neck	S2. Clavicular notch
R3. Tubercle	S3. Manubrium
R4. Body	S4. Body
R5. Sternal end (*attaches to costal cartilage*)	S5. Xiphoid process

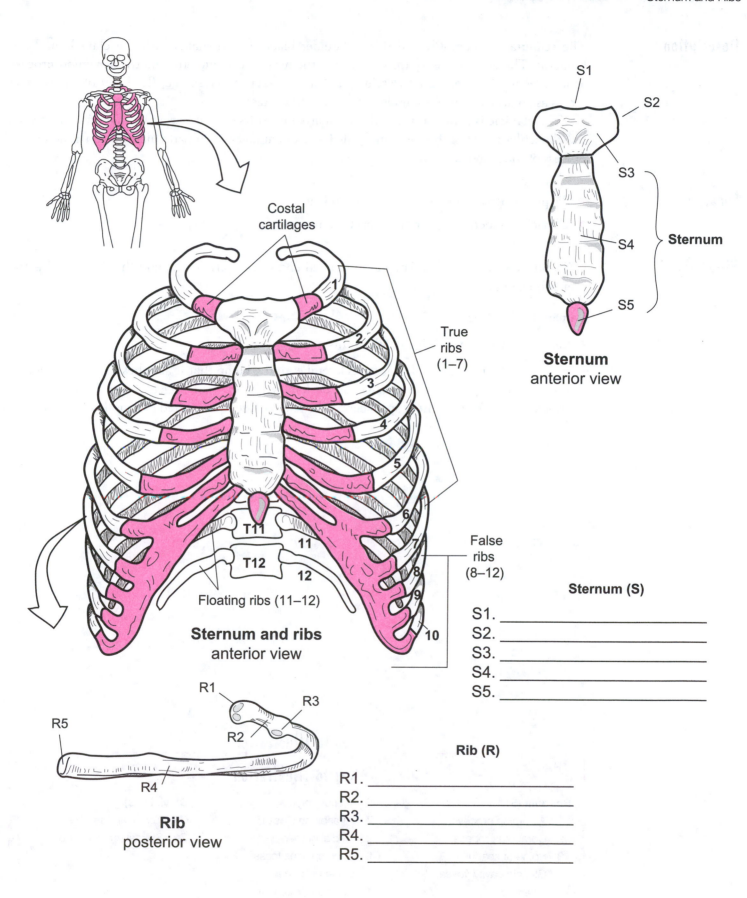

Costal cartilages

S1

S2

S3

S4

Sternum

S5

Sternum
anterior view

True ribs
(1–7)

False ribs
(8–12)

1

2

3

4

5

6

7

8

9

10

11

12

T11

T12

Floating ribs (11–12)

Sternum and ribs
anterior view

R1

R3

R2

R5

R4

Rib
posterior view

Sternum (S)

S1. _____

S2. _____

S3. _____

S4. _____

S5. _____

Rib (R)

R1. _____

R2. _____

R3. _____

R4. _____

R5. _____

Description

The **scapula** is commonly called the shoulder blade. It articulates with the clavicle and the humerus. The **glenoid cavity** (fossa) receives the head of the humerus and the **acromion process** articulates with the acromial (lateral) end of the clavicle. The large, flat **body** of the scapula serves as an attachment for muscles, tendons, and ligaments.

The **clavicle** is commonly called the collarbone. At its sternal (*medial*) end it articulates with the clavicular notch of the sternum, while its acromial (*lateral*) end articulates with the acromion process of the scapula.

Location

- **Scapula**—acromial region (shoulder) and back
- **Clavicle**—superior border of the pectoral region

Study Tips

- To position the clavicle: The more rounded end is the sternal end, and the flatter end is the acromial end.

Palpate (*feel by touch*):

- **Clavicle:** You can feel the details of this bone.

- Feel your clavicle, and follow it to its lateral end, the bump that it attaches to is the **acromion process.** The **spine** of the scapula can be felt by gliding your fingers along the back of the shoulder. The body of the scapula cannot be felt because it is covered by muscle.

Key to Illustration

Scapula (S)

1. Acromion process
2. Coracoid process
3. Spine of scapula
4. Glenoid cavity (*fossa*)
5. Lateral border
6. Inferior angle
7. Infraspinous fossa
8. Medial border
9. Supraspinous fossa
10. Superior border
11. Suprascapular notch

Clavicle (C)

C1. Acromial (*lateral*) end
C2. Sternal (*medial*) end

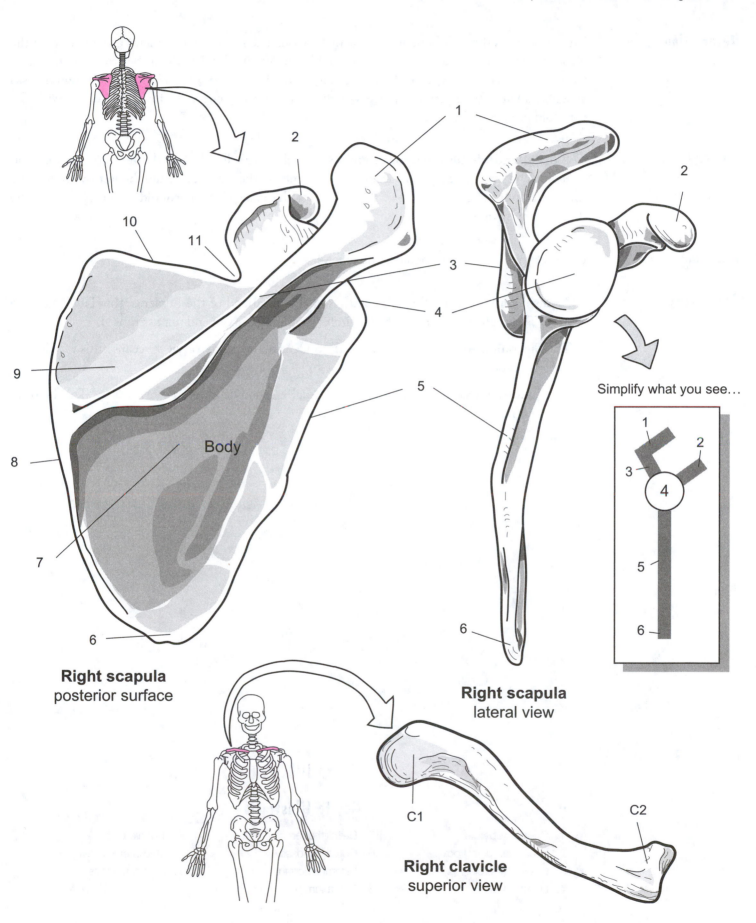

1

2

10

11

2

3

4

9

5

8

Body

7

6

Simplify what you see…

1

2

3

4

5

6

Right scapula
posterior surface

Right scapula
lateral view

C1

C2

Right clavicle
superior view

Description

The humerus is the only bone in the brachial region. The **head** of the humerus articulates with the glenoid cavity in the scapula to form the shoulder joint. At the distal end the **trochlea** articulates with the ulna, while the **capitulum** articulates with the head of the radius. The **olecranon fossa** articulates with the olecranon process of the ulna, and the **coronoid fossa** articulates with the coronoid process of the ulna.

Analogy

The **distal end of the humerus** resembles a **hitchhiker's hand**. The thumb is the medial epicondyle. The first two fingers adjacent to the thumb are the **trochlea**, while the last two fingers are the **capitulum**. The depression in the middle of the palm is the **coronoid fossa**. Note that the thumb always points medially.

Location

Brachial region (between shoulder and elbow)

Study Tips

- To distinguish between the two condyles at the distal end of the humerus: the **T**rochlea has a more **T**riangular shape, while the **C**apitulum is simply **C**urved and not as pointed.

- The **olecranon fossa** is a good landmark to identify the posterior view because it is the deepest depression on the bone.

- Note that the **head** of the humerus and **medial epicondyle** always point *medially*. This helps to distinguish a *left* humerus from a *right* humerus.

Key to Illustration

1. Greater tubercle	5. Deltoid tuberosity	9. Trochlea
2. Intertubercular groove	6. Coronoid fossa	10. Medial epicondyle
3. Lesser tubercle	7. Lateral epicondyle	11. Olecranon fossa
4. Head	8. Capitulum	

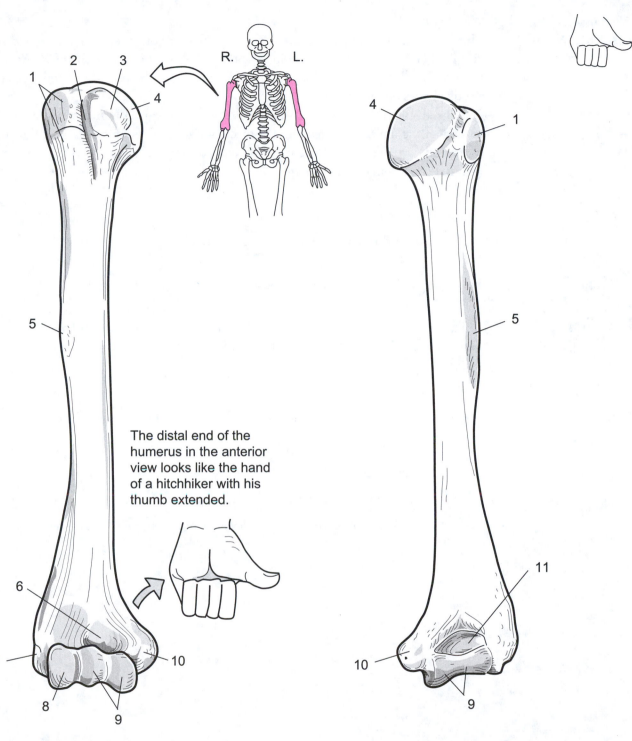

The distal end of the humerus in the anterior view looks like the hand of a hitchhiker with his thumb extended.

Right humerus
anterior view

Right humerus
posterior view

1. _____
2. _____
3. _____
4. _____
5. _____
6. _____

7. _____
8. _____
9. _____
10. _____
11. _____

107

Description

The radius and the ulna articulate with each other at both their proximal and distal ends. At the proximal end, the head of the radius pivots on the radial notch of the ulna. At the distal end the head of the ulna joins the ulnar notch of the radius. A fibrous sheet of connective tissue connects the **diaphyses** of both bones.

The distal end of the radius articulates with the **carpal bones** in the wrist. The **olecranon process** of the ulna articulates with the olecranon fossa of the humerus, while the **coronoid process** of the ulna articulates with the coronoid fossa of the humerus. The head of the radius articulates with the capitulum of the humerus.

Analogy

- The **ulna** resembles a **crescent wrench**.

- The **head of the radius** resembles a **warped hockey puck**.

Location

The radius and ulna are located in the antebrachial region (forearm)

Study Tips

- The proximal end of the ulna has a "U" shaped structure called the trochlear notch that identifies it as the **U**lna because "U" is the first letter in the word "Ulna."

- The head of the **R**adius is **R**ounded. Use the alliteration, "**R**ounded **R**adius' to distinguish the radius from the ulna.

- Notches and other structures are sometimes named after the bones with which they articulate:

  e.g. —radial notch of ulna

    —ulnar notch of radius

    —olecranon *process* of the ulna hooks into the olecranon *fossa* of the humerus

    —coronoid *process* of the ulna hooks into the coronoid *fossa* of the humerus

Key to Illustration

Radius	**Ulna**	
1. Neck	5. Coronoid process	9. Head of ulna
2. Radial tuberosity	6. Olecranon process	10. Ulnar styloid process
3. Radial styloid process	7. Trochlear notch	11. Radial notch of ulna
4. Ulnar notch of radius	8. Ulnar tuberosity	
(not fully visible in diagram)		

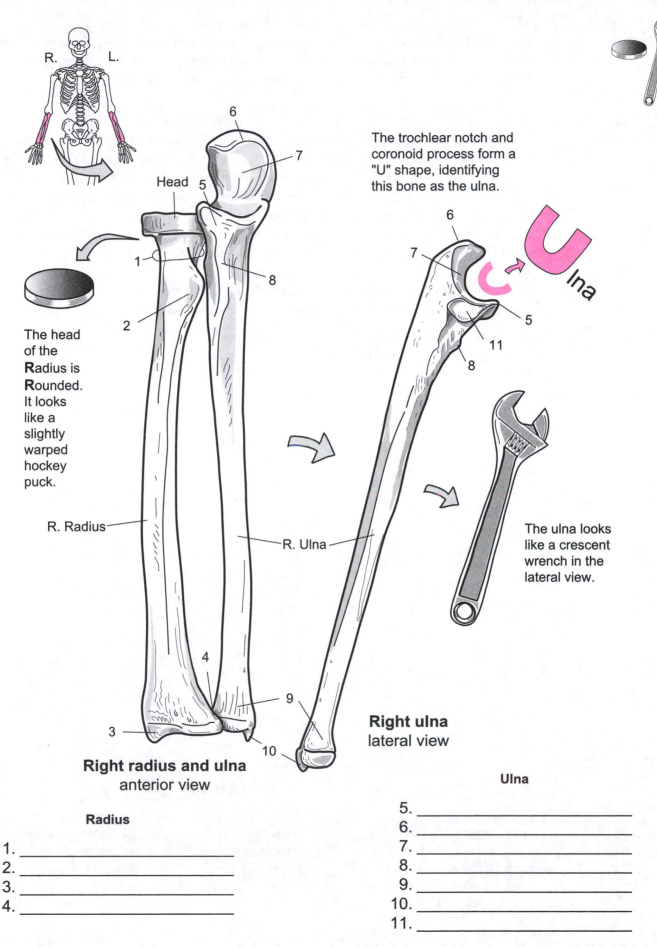

The head of the **R**adius is **R**ounded. It looks like a slightly warped hockey puck.

Head

The trochlear notch and coronoid process form a "U" shape, identifying this bone as the ulna.

Ulna

The ulna looks like a crescent wrench in the lateral view.

R. Radius

R. Ulna

Right ulna
lateral view

Right radius and ulna
anterior view

Radius

1. _____
2. _____
3. _____
4. _____

Ulna

5. _____
6. _____
7. _____
8. _____
9. _____
10. _____
11. _____

Description

Each hand contains a total of 27 bones and is divided into three groups of bones: **carpals** (8), **metacarpals** (5), and **phalanges** (14). The carpals are the small bones of the wrist. The palm of the hand contains the metacarpals, and the phalanges are located in the fingers or digits.

Study Tips

The most difficult part of the hand to learn is the *carpals*. These eight bones are small and clustered together in the wrist. Simplify it by viewing them as two equal rows of bones stacked one on top of the other. Each row has four bones in it. The first row is nearer the radius and ulna. The second row is next to the metacarpals. Beginning on the thumb side of the first row, the proper order is: **scaphoid, lunate, triangular, pisiform.** Following the same pattern, the proper order for the second row is: **trapezium, trapezoid, capitate, hamate.**

The *hamate (hamatum,* hooked*) bone* has a hook-like process on it (*visible only in the anterior view*). You can better recall this bone because the words **h**amate and **h**ook both begin with the letter "**h**".

The *thumb* has only two phalanges instead of three. It has a proximal and distal phalanx but lacks a middle phalanx. This is the same pattern as the great toe of the foot.

● On many plastic models, the triangular (*triquetrum*) and pisiform bones are often fused together to appear as one bone. This causes some confusion for students.

Key to Illustration

Carpals
1. Scaphoid
2. Lunate
3. Triangular (*Triquetrum*)
4. Pisiform

5. Trapezium
6. Trapezoid
7. Capitate
8. Hamate

Phalanges
 9. Proximal phalanx
10. Middle phalanx
11. Distal phalanx

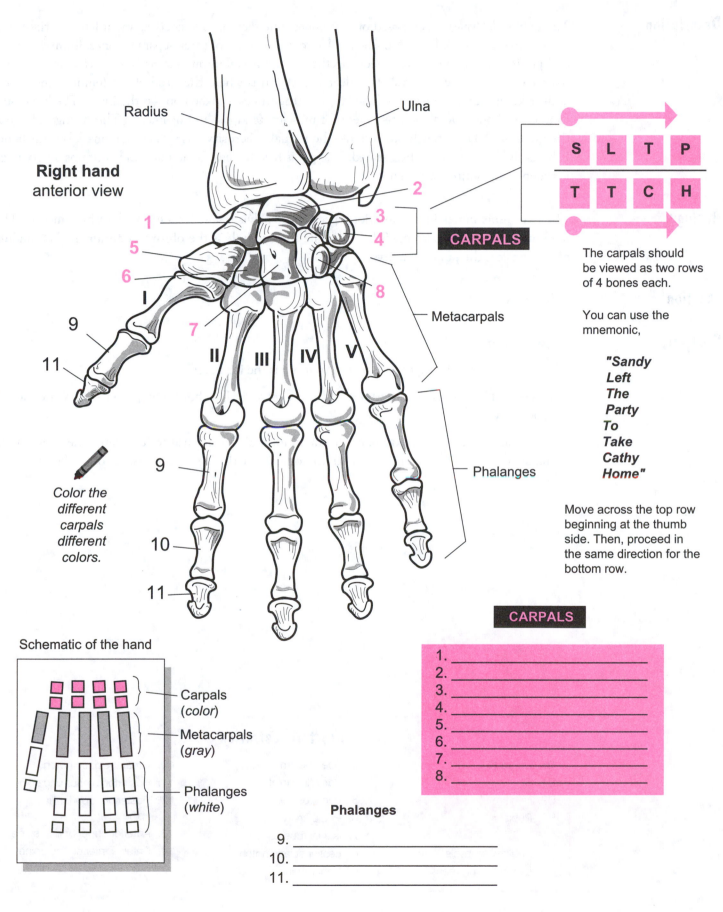

Right hand
anterior view

Radius

Ulna

CARPALS

Metacarpals

Phalanges

I II III IV V

Color the different carpals different colors.

Schematic of the hand

Carpals (*color*)

Metacarpals (*gray*)

Phalanges (*white*)

S	L	T	P
T	T	C	H

CARPALS

The carpals should be viewed as two rows of 4 bones each.

You can use the mnemonic,

*"Sandy
Left
The
Party
To
Take
Cathy
Home"*

Move across the top row beginning at the thumb side. Then, proceed in the same direction for the bottom row.

CARPALS

1. _____
2. _____
3. _____
4. _____
5. _____
6. _____
7. _____
8. _____

Phalanges

9. _____
10. _____
11. _____

Description

The **pelvis** (*hipbone*) is composed of four bones: one sacrum, one coccyx, and a left and right ossa coxae (sing. *os coxae*). Each os coxae results from the fusion of three separate bones: **ilium**, **ischium**, and **pubis** bones. These three meet together in the **acetabulum** (*means "cup of vinegar"*) to form what looks like a "peace sign." This deep depression is where the head of the femur forms a ball-and-socket joint with the pelvis. The largest of the three os coxae bones is the ilium. The broad surface of this bone is a major attachment for muscles, tendons, and ligaments. The strongest bone is the ischium. A large, rough projection—the **ischial tuberosity**—is a major landmark on this bone. It bears the body weight when seated. The pubis has the least bone mass and contains a very large foramen—the **obturator foramen**.

Analogy

The **two pubis bones** in the anterior aspect of the pelvis resembles a **mask** worn by someone. The mask itself represents both pubis bones, while the **eye hole** is the **obturator foramen**. The **middle of the mask** is the **pubic symphysis**.

Location

Pelvic region

Study Tips

Palpate (*feel by touch*):

- The curve along the superior aspect of your hip is the **iliac crest**.

- Place your hand on your buttock and push inward. The bony bump you feel is the **ischial tuberosity**.

- Good landmarks on a coxal bone: **acetabulum** and the **obturator foramen**. The acetabulum marks the lateral aspect, and the obturator foramen marks the anterior aspect of the bone.

Key to Illustration

Os Coxae (B)
B1. Ilium
B2. Ischium
B3. Pubis

1. Pubic symphysis
2. Obturator foramen

3. Acetabulum
4. Sacroiliac joint
5. Iliac crest
6. Pubic crest
7. Ischial tuberosity
8. Lesser sciatic notch
9. Ischial spine

10. Greater sciatic notch
11. Posterior inferior iliac spine
12. Posterior superior iliac spine
13. Anterior superior iliac spine
14. Anterior inferior iliac spine

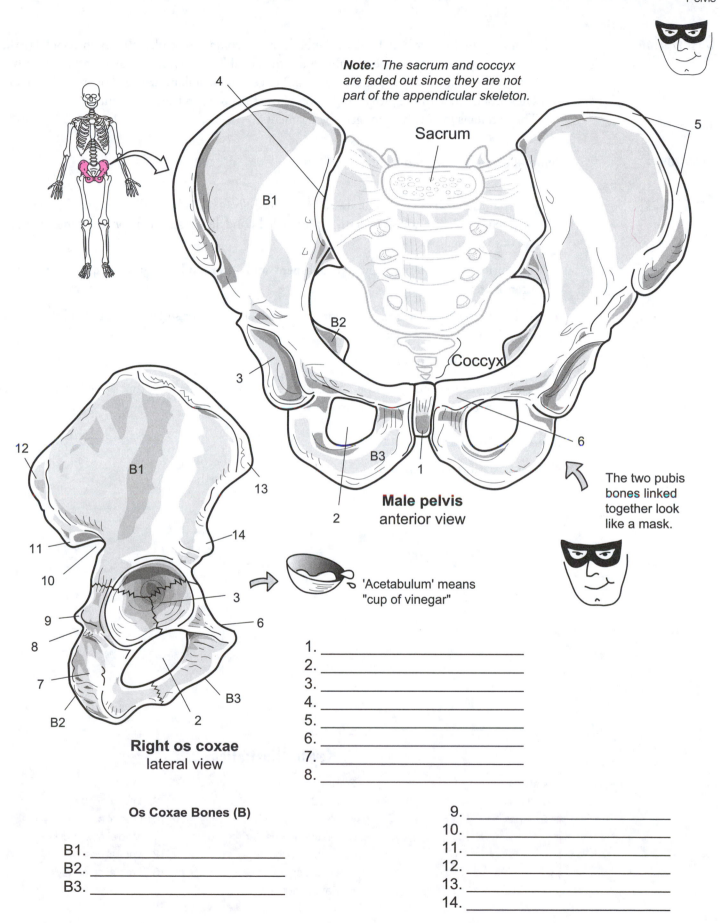

Note: *The sacrum and coccyx are faded out since they are not part of the appendicular skeleton.*

Sacrum

Coccyx

Male pelvis
anterior view

The two pubis bones linked together look like a mask.

'Acetabulum' means "cup of vinegar"

Right os coxae
lateral view

Os Coxae Bones (B)

B1. _____
B2. _____
B3. _____

1. _____
2. _____
3. _____
4. _____
5. _____
6. _____
7. _____
8. _____

9. _____
10. _____
11. _____
12. _____
13. _____
14. _____

113

Description

The **femur** is the longest long bone in the body and is commonly called the thigh bone. The **head** of the femur articulates with the acetabulum of the coxal bone to form the ball-and-socket joint in the hip. At the distal end, the patella (kneecap) covers the **patellar surface**. The **medial** and **lateral condyles** articulate with the proximal portion of the tibia to form the knee joint.

The **patella** is loosely held in place by ligaments.

Location

Femoral region (between hip and knee)

Study Tips

- Note that the **head** of the femur always points *medially*. This helps to distinguish a *left* femur from a *right* femur. It also helps to identify the medial condyle because it is on the same side as the head of the humerus.

- To identify the posterior view: **linea aspera** and **intercondylar groove** appear only in the posterior view.

- To identify the anterior from posterior in patella: The anterior surface is rough and the posterior surface is much more smooth.

- **LAP** = **L**inea **A**spera is **P**osterior

- For distinguishing the apex from the base of the patella, remember that apex is a general term meaning the *pointed tip* of a structure.

Palpate (*feel by touch*):

- You can easily feel your patella.

Key to Illustration

Patella
1. Base
2. Apex

Femur
3. Head
4. Neck
5. Greater trochanter
6. Lesser trochanter
7. Linea aspera
8. Patellar surface
9. Intercondylar groove
10. Medial condyle
11. Medial epicondyle
12. Lateral condyle
13. Lateral epicondyle

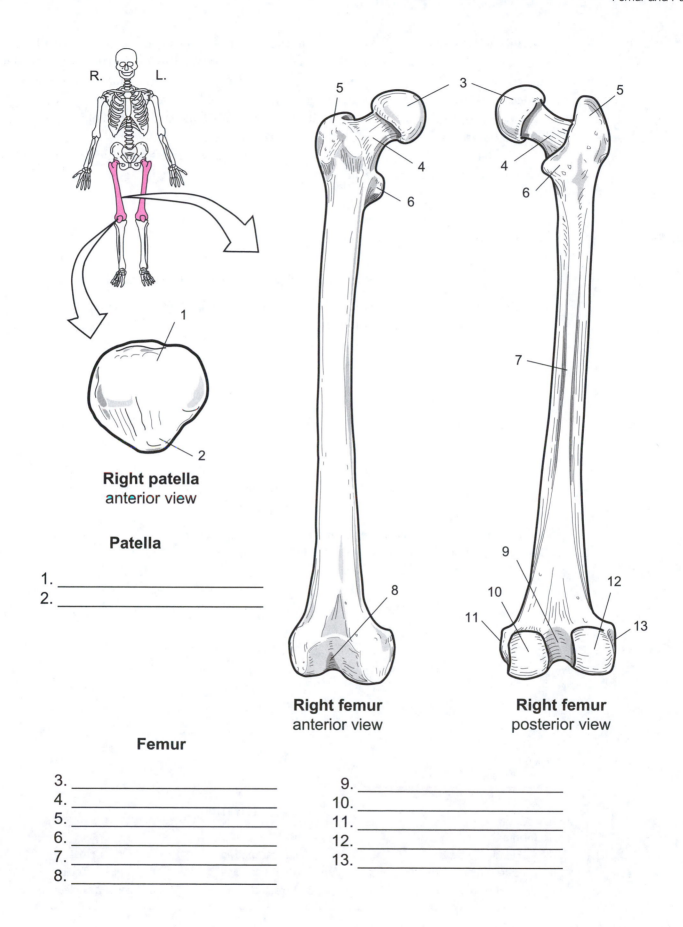

Right patella
anterior view

Right femur
anterior view

Right femur
posterior view

Patella

1. _____
2. _____

Femur

3. _____
4. _____
5. _____
6. _____
7. _____
8. _____

9. _____
10. _____
11. _____
12. _____
13. _____

Description

The **tibia** and the **fibula** are the two bones in the leg between the knee and ankle. The **tibia** is commonly called the shinbone and is the larger of the two bones. At its **proximal** surface it articulates with the **distal** end of the femur to form the knee joint. At its distal end, the **inferior articular surface** articulates with the talus in the foot to form the ankle joint. The **anterior margin** is a ridge that runs along the shaft of the bone on the anterior surface. The **medial malleolus** is a large process that stabilizes the ankle joint.

The **fibula** is the smaller bone. The **head** of the fibula articulates with the proximal end of the tibia. The **lateral malleolus** articulates with the distal end of the tibia and with the talus in the ankle.

Location

Crural region (between knee and ankle)

Study Tips

- To distinguish between the **tibia** and the **fibula**: The tibia is the larger of the two bones.

- To distinguish between the two different ends of the fibula: The **lateral malleolus** is more tapered and triangular in shape and is located near the ankle; the **head** of the **fibula** is more rounded in shape and is located near the knee.

Palpate (*feel by touch*):

- Touch the large bump on the medial side of your ankle. This is the **medial malleolus** of the tibia.

- Now touch the bony bump on the lateral side of your ankle. This is the **lateral malleolus** of the **fibula**.

- Feel your patella (*kneecap*), and slide your hand straight down toward your ankle. The first small bump you feel below the knee is the **tibial tuberosity** of the tibia. The soft spot between your knee and the tibial tuberosity is the **patellar ligament**.

- Feel the long ridge of bone beginning below the **tibial tuberosity** and running down toward the ankle. This is the **anterior crest** of the tibia.

Key to Illustration

Fibula	Tibia	
1. Head	4. Lateral condyle	7. Anterior crest
2. Lateral malleolus	5. Medial condyle	8. Medial malleolus
3. Inferior articular surface	6. Tibial tuberosity	9. Inferior articular surface

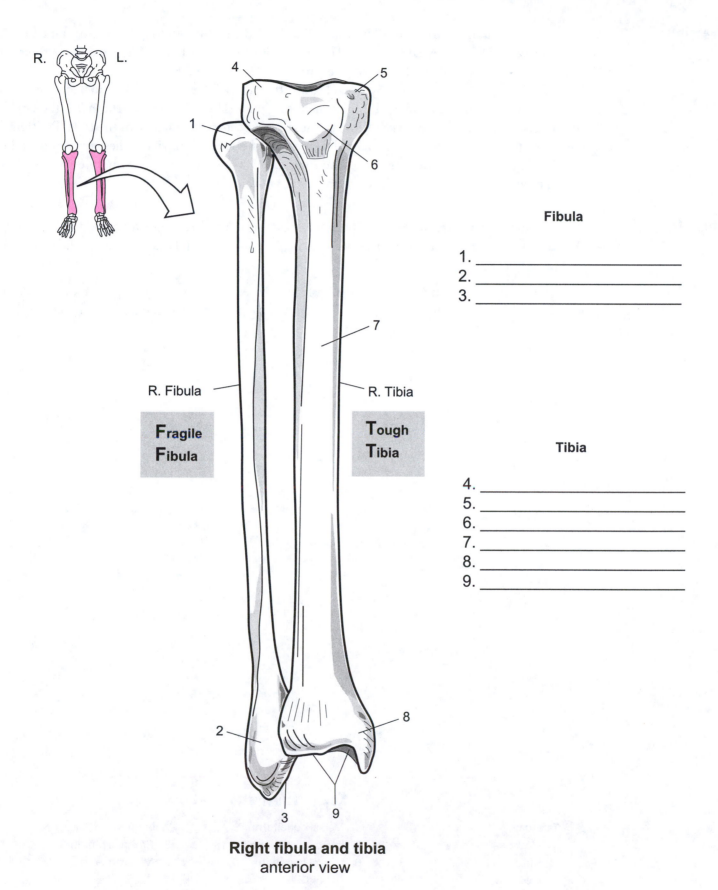

Fibula

1. _____
2. _____
3. _____

Tibia

4. _____
5. _____
6. _____
7. _____
8. _____
9. _____

R. Fibula

Fragile Fibula

R. Tibia

Tough Tibia

Right fibula and tibia
anterior view

Description

Each foot contains a total of 26 bones and is divided into three groups of bones: **tarsals** (7), **metatarsals** (5), and **phalanges** (14). The tarsals are the ankle bones. The middle of the foot contains the metatarsals, and the phalanges are located in the toes or digits.

The most difficult part of the foot to learn is the *tarsals*. These seven bones include the **talus, calcaneus, navicular, cuboid,** and three **cuneiform** bones. The **calcaneus** (*heel bone*) is the largest bone in this group. The talus is curved on its superior aspect to articulate with the tibia. Positioned between the calcaneus and the metatarsals is the **cuboid** bone. The **navicular** lies between the talus and cuneiforms. The **cuneiform** bones are the smallest bones in this group and are identified by their position within the foot (*medial, intermediate,* and *lateral*).

Study Tips

The **talus bone** articulates with the tibia. It is also the tallest bone when the foot is viewed laterally. You can easily recall this because the words *T*alus and *T*ibia and *T*allest all begin with the letter "*t*."

The **great toe** has only two phalanges instead of three. It has a proximal phalanx and a distal phalanx but lacks a middle phalanx. This is the same pattern as the thumb of the hand.

Key to Illustration

Tarsals		Phalanges
1. Calcaneus	4. Medial **cuneiform**	8. Proximal phalanx
2. Talus	5. Intermediate **cuneiform**	9. Middle phalanx
3. Navicular	6. Lateral **cuneiform**	10. Distal phalanx
	7. Cuboid	

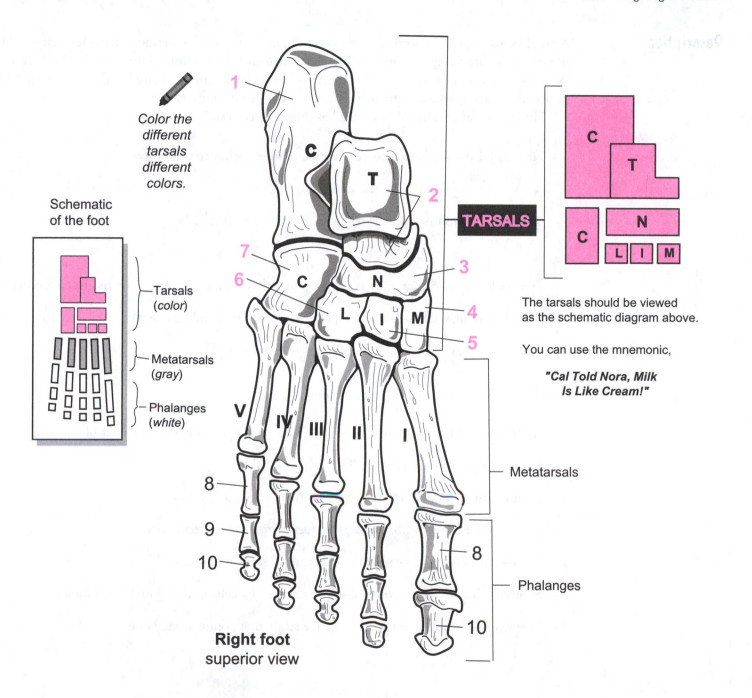

Color the different tarsals different colors.

Schematic of the foot

Tarsals (*color*)

Metatarsals (*gray*)

Phalanges (*white*)

Right foot
superior view

TARSALS

The tarsals should be viewed as the schematic diagram above.

You can use the mnemonic,

"Cal Told Nora, Milk Is Like Cream!"

Metatarsals

Phalanges

TARSALS

1. _____
2. _____
3. _____
4. _____
5. _____
6. _____
7. _____

Phalanges

8. _____
9. _____
10. _____

Description

Articulations (*joints*) are formed when two or more bones meet. Common examples that quickly come to mind are freely moving joints like the knee and elbow joints. Sutures in the skull are also joints but are often overlooked since they have no movement associated with them. Joints are classified according to either their structure or their degree of movement.

The structural classification system has three categories:

- **Fibrous joints**—only **fibrous connective tissues** anchor the joint together

  ex: sutures in the skull

- **Cartilaginous joints**—only **cartilage** anchors the joint together

  ex: pubic symphysis

- **Synovial joints**—structurally complex joints that contain a **variety of tissues/structures**

  such as articular cartilage, joint capsule, synovial fluid, and ligaments.

  ex: knee, shoulder, elbow, and hip joints

Here are the key structures in any **synovial joint**:

- **Articular cartilage**—smooth layer of hyaline cartilage that covers the ends of long bones

- **Joint cavity**—potential space filled with synovial fluid

- **Joint capsule**—double layered structure that surrounds the joint

  —outer layer = **fibrous capsule** (*dense irregular connective tissue*)

  —inner layer = **synovial membrane** (*loose connective tissue*)

- **Synovial fluid**—viscous, oily substance secreted by cells in the synovial membrane

- **Ligaments**—bands of fibrous connective tissue that connect one bone to another

Key to Illustration

1. Periosteum	5. Fibrous capsule	8. Joint cavity (*filled with synovial fluid*)
2. Medullary cavity	6. Synovial membrane	9. Articular cartilage
3. Yellow marrow	7. Joint capsule	
4. Ligament		

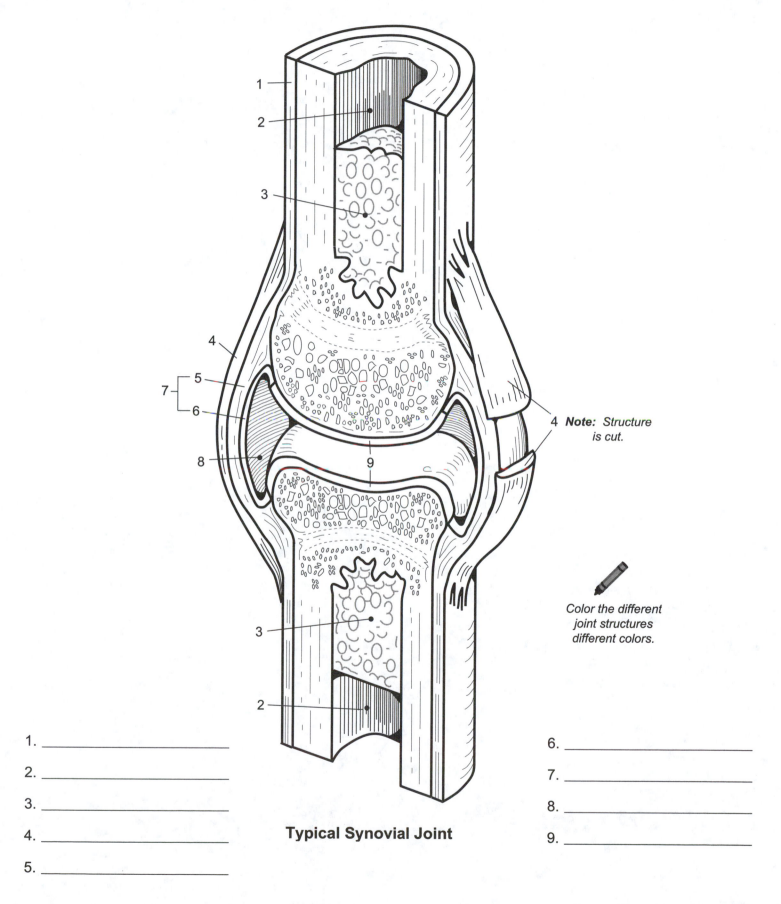

Typical Synovial Joint

Note: Structure is cut.

Color the different joint structures different colors.

1. _____

2. _____

3. _____

4. _____

5. _____

6. _____

7. _____

8. _____

9. _____

Notes

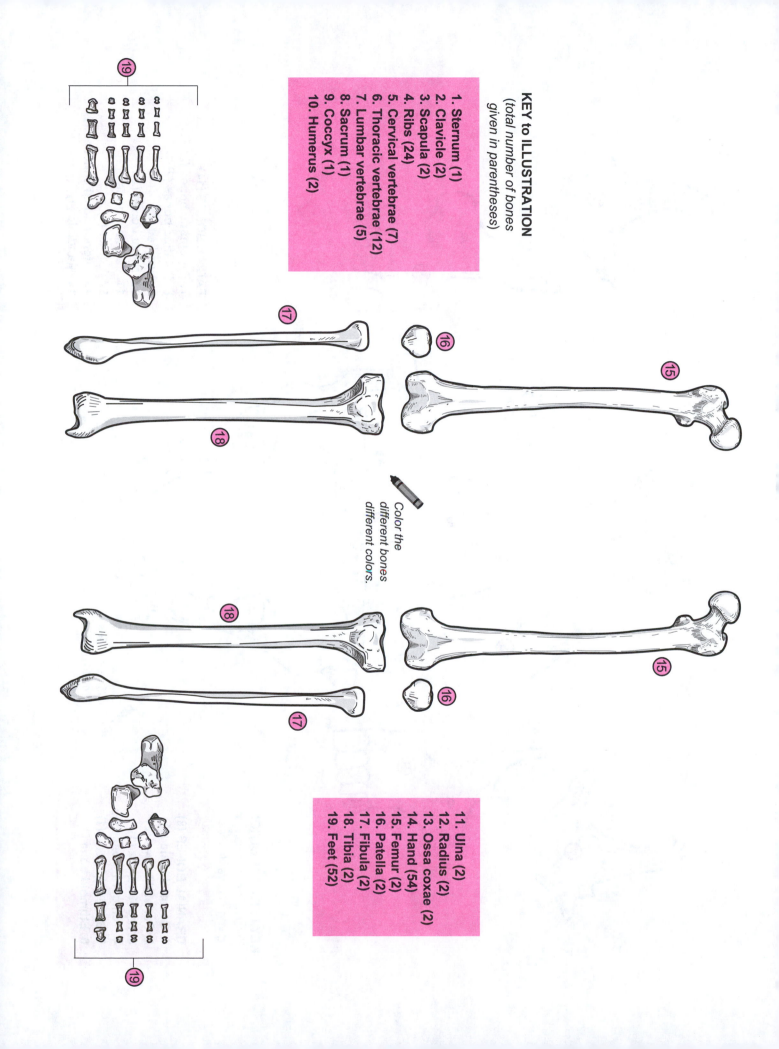

Color the
different bones
different colors.

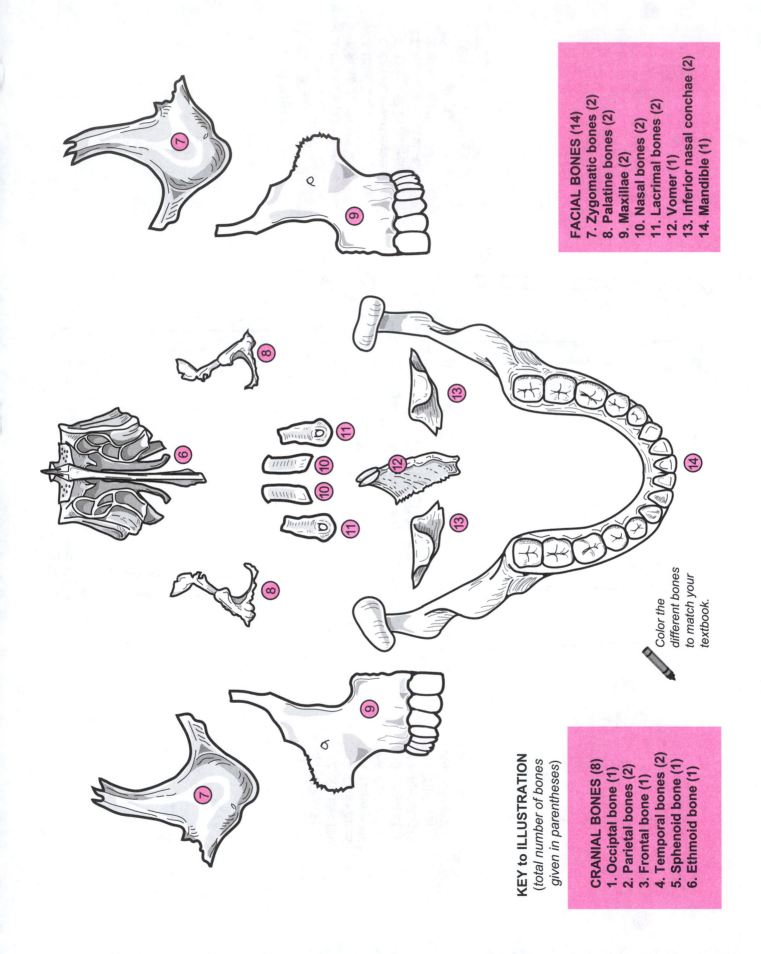

Color the different bones to match your textbook.

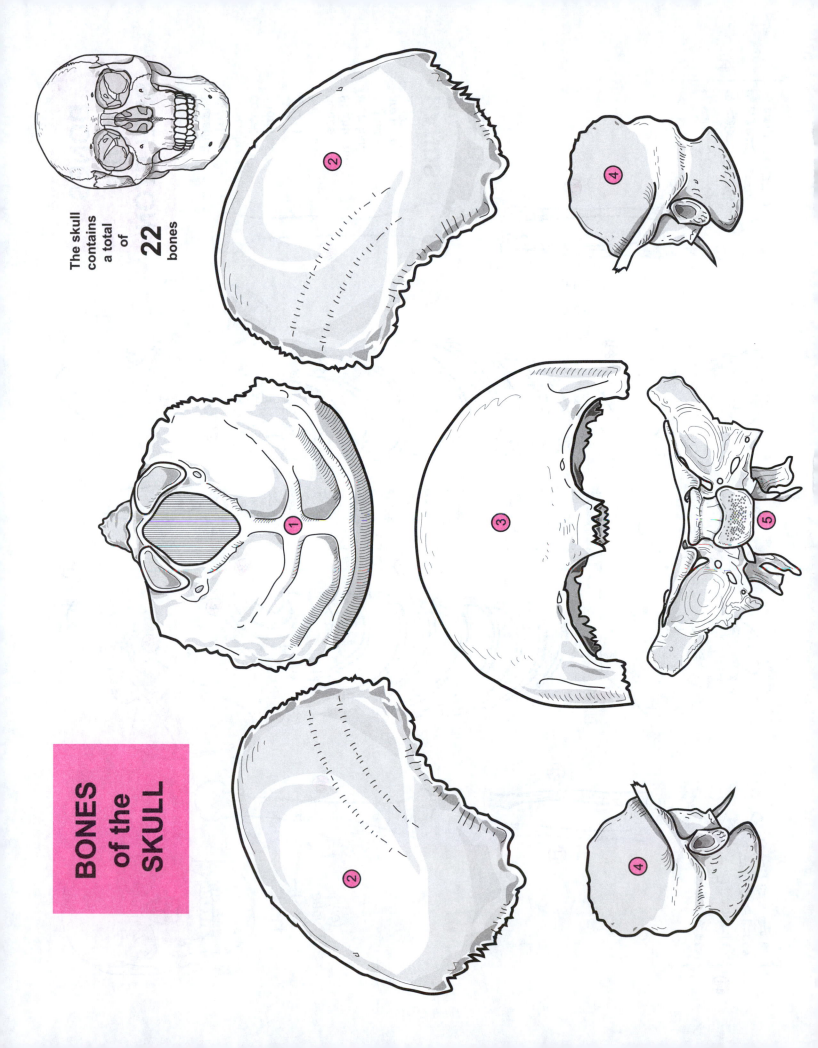

The skull contains a total of **22** bones

BONES of the SKULL

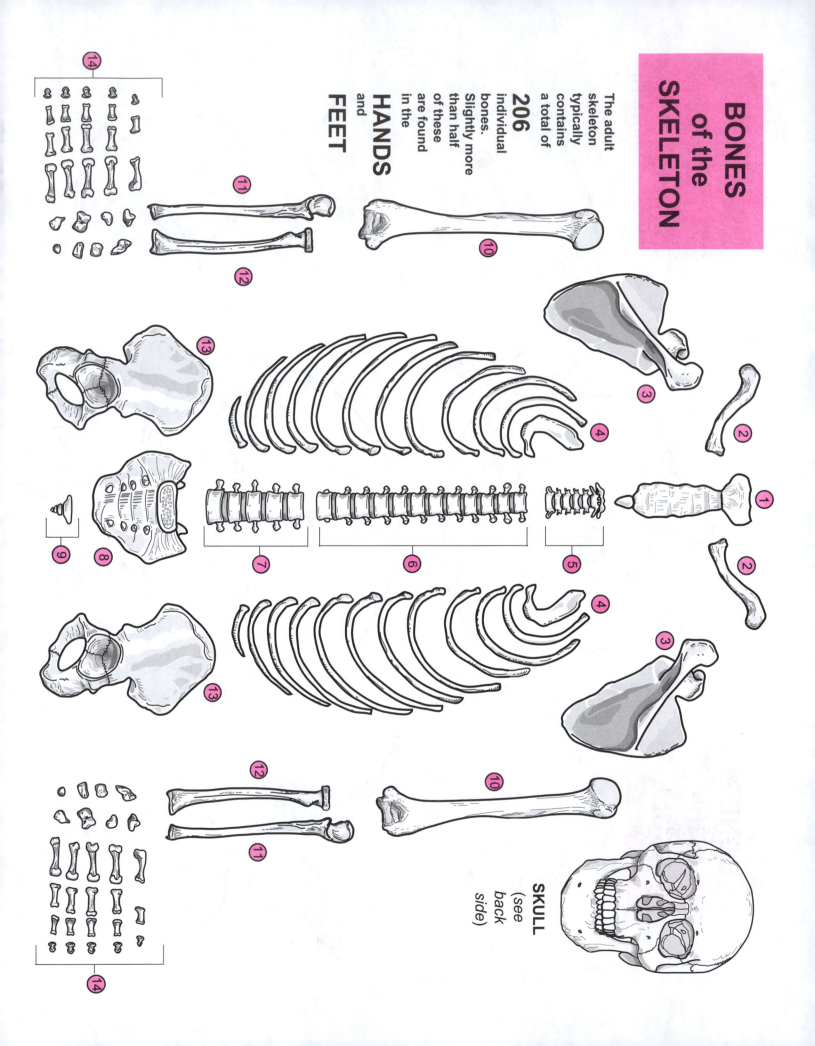

BONES
of the
SKELETON

The adult skeleton typically contains a total of **206** individual bones. Slightly more than half of these are found in the

HANDS
and
FEET

SKULL
(see back side)

Muscular System

Description

A whole skeletal muscle is packaged like a series of tubes within other tubes. A muscle first is subdivided into a bundle of long tubes called **fascicles.** Each fascicle is a bundle of skeletal muscle cells or fibers. Each skeletal muscle cell is a bundle of **myofibrils**. Each myofibril is composed primarily of two different protein filaments: **actin** and **myosin**. Because of their difference in size, myosin is called the thick filament and actin is called the thin filament. These proteins are part of a repeated unit called a **sarcomere**, which is the structural and functional unit for muscle contraction. The ends of a sarcomere are defined by the **Z-lines**, which are made of protein.

All of these different bundles of tube-like structures are held together with connective tissue. The **epimysium** is a tough, fibrous, connective tissue that completely surrounds the outside of a whole muscle. Within the whole muscle, the **perimysium** fills the space between fascicles. Surrounding each skeletal muscle cell is another connective tissue called the **endomysium**, which mainly serves to bind one skeletal muscle cell to another.

Analogies

Three analogies are given for structures at the level of the sarcomere.

1. An **actin** or thin filament is compared to a **double-stranded chain of pearls**. Each **pearl** is equivalent to **one molecule of actin**. (Note that this analogy does not include the troponin and tropomyosin proteins.)

2. The myosin filament has **heads** (cross bridges) branching off from it, which later attach to actin during muscle contraction. From the lateral view, these heads appear angled like the **tail feathers in an arrow**.

3. The **myosin heads** attach to actin and pull on it with a regular movement. To visualize this movement, imagine the **heads moving** like a **boat rower's oars**. Unlike the oar movements, however, the heads do not all move at the same time.

Location

All skeletal muscles in the human body (more than 600 in all)

Function

Contraction

Key to Illustration

1. Whole muscle (biceps brachii from the illustration)
2. Fascicle
3. Skeletal muscle cell (fiber)
4. Myofibril
5. Epimysium
6. Perimysium
7. Endomysium
8. Myosin heads
9. Myosin (*thick*) filaments
10. Actin (*thin*) filaments
11. Z-line

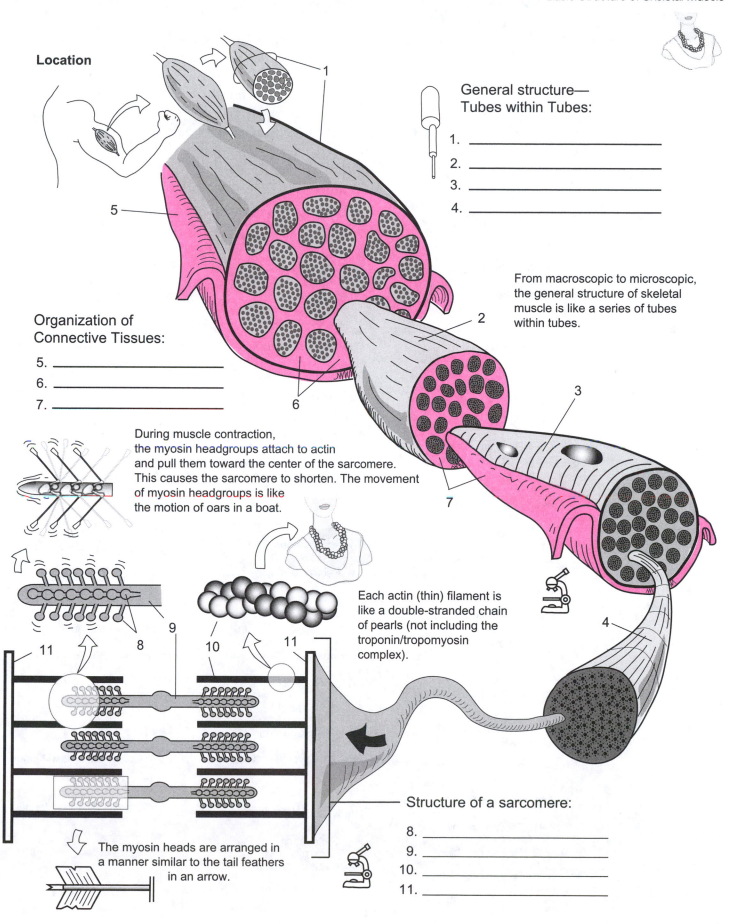

Location

General structure— Tubes within Tubes:

1. _____
2. _____
3. _____
4. _____

From macroscopic to microscopic, the general structure of skeletal muscle is like a series of tubes within tubes.

Organization of Connective Tissues:

5. _____
6. _____
7. _____

During muscle contraction, the myosin headgroups attach to actin and pull them toward the center of the sarcomere. This causes the sarcomere to shorten. The movement of myosin headgroups is like the motion of oars in a boat.

Each actin (thin) filament is like a double-stranded chain of pearls (not including the troponin/tropomyosin complex).

The myosin heads are arranged in a manner similar to the tail feathers in an arrow.

Structure of a sarcomere:

8. _____
9. _____
10. _____
11. _____

125

Key to Illustration

Facial Muscles (F)

F1. Frontalis

F2. Temporalis

F3. Platysma

F4. Orbicularis oculi

F5. Masseter

F6. Orbicularis oris

F7. Occipitalis

Neck Muscles (N)

N1. Sternohyoid

N2. Sternocleidomastoid

Shoulder Muscles (S)

S1. Trapezius

S2. Deltoid

S3. Infraspinatus

S4. Teres minor

S5. Teres major

Arm Muscles (AR)

AR1. Triceps brachii

AR2. Biceps brachii

AR3. Brachialis

Forearm Muscles (FO)

FO1. Pronator teres

FO2. Brachioradialis

FO3. Flexor carpi radialis

FO4. Palmaris longus

FO5. Extensor carpi radialis longus

FO6. Flexor carpi ulnaris

FO7. Extensor digitorum

FO8. Extensor carpi ulnaris

Thorax Muscles (T)

T1. Pectoralis major

T2. Serratus anterior

Abdominal Muscles (AB)

AB1. Rectus abdominis

AB2. External oblique

AB3. Internal oblique

AB4. Transverse abdominis

Hip Muscles (H)

H1. Gluteus medius

H2. Gluteus maximus

Back (B)

B1. Latissimus dorsi

Pelvis/Thigh (TH)

TH1. Tensor fasciae latae

TH2. Iliopsoas

TH3. Sartorius

TH4. Pectineus

TH5. Adductor longus

TH6. Gracilis

TH7. Rectus femoris

TH8. Vastus lateralis

TH9. Vastus medialis

TH10. Adductor magnus

TH11. Biceps femoris

TH12. Semitendinosus

TH13. Semimembranosus

Leg (L)

L1. Fibularis (*peroneus*) longus

L2. Tibialis anterior

L3. Extensor digitorum longus

L4. Gastrocnemius

L5. Soleus

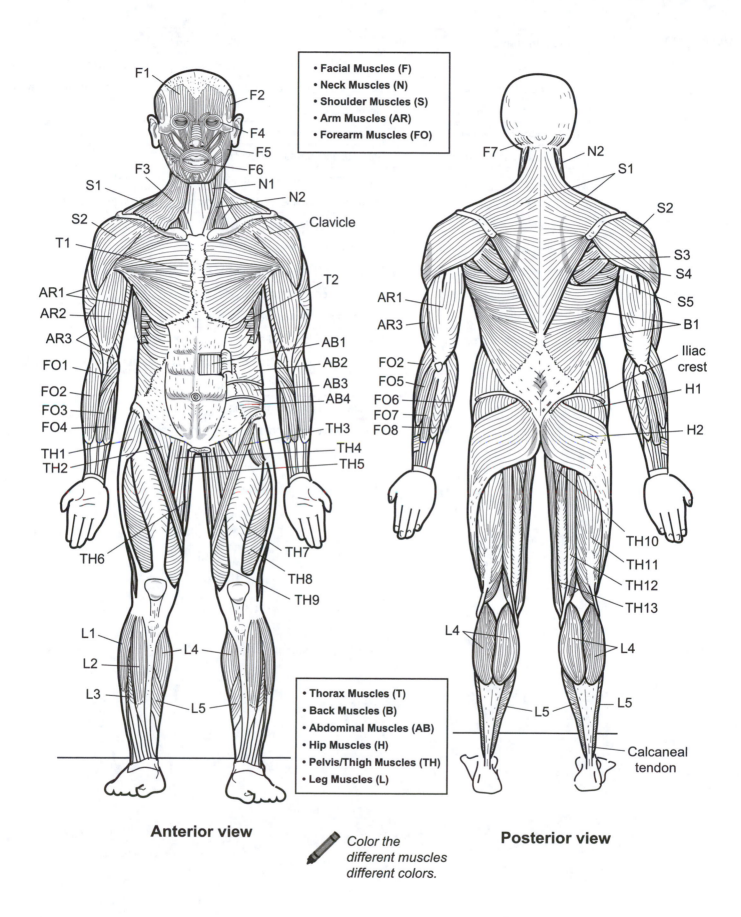

- **Facial Muscles (F)**
- **Neck Muscles (N)**
- **Shoulder Muscles (S)**
- **Arm Muscles (AR)**
- **Forearm Muscles (FO)**

Anterior view

Color the different muscles different colors.

Posterior view

- **Thorax Muscles (T)**
- **Back Muscles (B)**
- **Abdominal Muscles (AB)**
- **Hip Muscles (H)**
- **Pelvis/Thigh Muscles (TH)**
- **Leg Muscles (L)**

Description

Muscle Name	Action
1. Frontalis	Wrinkles skin of forehead, elevates eyebrows; draws scalp anteriorly
2. Orbicularis oculi	Depresses upper eyelid, elevates lower eyelid, tightens skin around eyes
3. Levator labii superioris	Elevates upper lip
4. Zygomaticus major	Retracts and elevates corner of mouth
5. Zygomaticus minor	Retracts and elevates upper lip
6. Risorius	Draws corner of mouth to the side
7. Platysma	Depresses mandible, pulls lower lip back and down *(as in pouting)*
8. Orbicularis oris	Closes, protrudes, and purses lips *(kissing muscle)*
9. Depressor labii inferioris	Depresses lower lip
10. Mentalis	Elevates and protrudes lower lip
11. Depressor anguli oris	Depresses corner of mouth
12. Buccinator	Compresses cheek inward *(as in whistling)*
13. Corrugator supercilii	Pulls skin inferiorly and anteriorly; wrinkles brow
14. Nasalis	Compresses bridge, depresses tip of nose; elevates corners of nostrils
15. Masseter	Elevates mandible and closes jaw
16. Temporalis	Elevates and retracts mandible
17. Corrugator supercilii	Pulls skin inferiorly and anteriorly; wrinkles brow

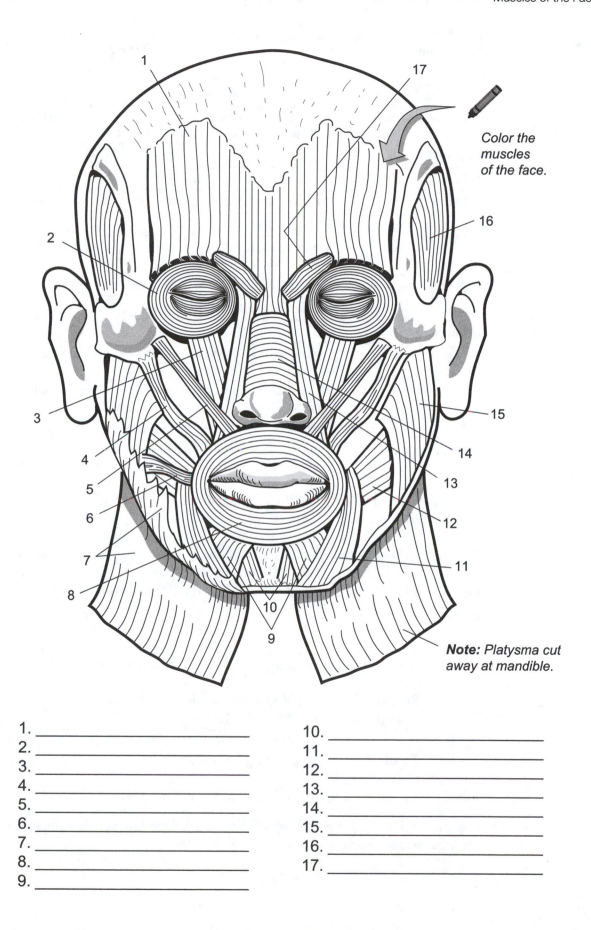

Color the
muscles
of the face.

Note: Platysma cut
away at mandible.

1. _____
2. _____
3. _____
4. _____
5. _____
6. _____
7. _____
8. _____
9. _____

10. _____
11. _____
12. _____
13. _____
14. _____
15. _____
16. _____
17. _____

Description

Muscle Name	Action
Neck (N)	
N1. Platysma	Depresses mandible; pulls lower lip back and down (as in pouting)
N2. Sternocleidomastoid	Simultaneous contraction of both heads: flexes neck. Individual action of each head: rotates head to shoulder on opposite side
Shoulder (S)	
S1. Trapezius	Extends neck; retracts scapula
S2. Deltoid	Flexes, extends, abducts; medially and laterally rotates arm
Thorax (T)	
T1. Pectoralis major	Primary muscle of arm flexion; adducts and medially rotates arm; with arm fixed, pulls chest forward (as in a forced inspiration)
T2. Pectoralis minor	Depresses and protracts scapula, elevates ribs
T3. Serratus anterior	Abducts and stabilizes scapula
Abdomen (A)	
A1. External oblique	Compresses anterior abdominal wall; flexes trunk; rotates trunk; depresses lower ribs
A2. Internal oblique	Compresses anterior abdominal wall; flexes trunk; rotates trunk; depresses lower ribs
A3. Transverse abdominis	Compresses anterior abdominal contents
A4. Rectus abdominis	Compresses anterior abdominal wall; flexes trunk

Analogies

- In the anterior, superficial view, the anterior portion of the **serratus anterior** looks like the blade of a **serrated knife**.

- The three abdominal muscles on the lateral surface of the abdomen—*external oblique*, *internal oblique*, and *transverse abdominis*—are layered on top of each other like a ham sandwich. The **internal oblique** is the **ham** (*in the middle*), and the **external oblique** and **transverse abdominis** are the layers of **bread**.

Study Tips

- Mnemonic for the 4 abdominal muscles: *"Really? Everything is terrible?"*

Really?	**R**ectus abdominis
Everything	**E**xternal oblique
Is	**I**nternal oblique
Terrible?	**T**ransverse abdominis

- *Rectus* means *straight*—this indicates the muscle fiber direction.

- Imagine you have your hands in your front pockets. Your fingers follow the fiber direction of the external oblique muscle.

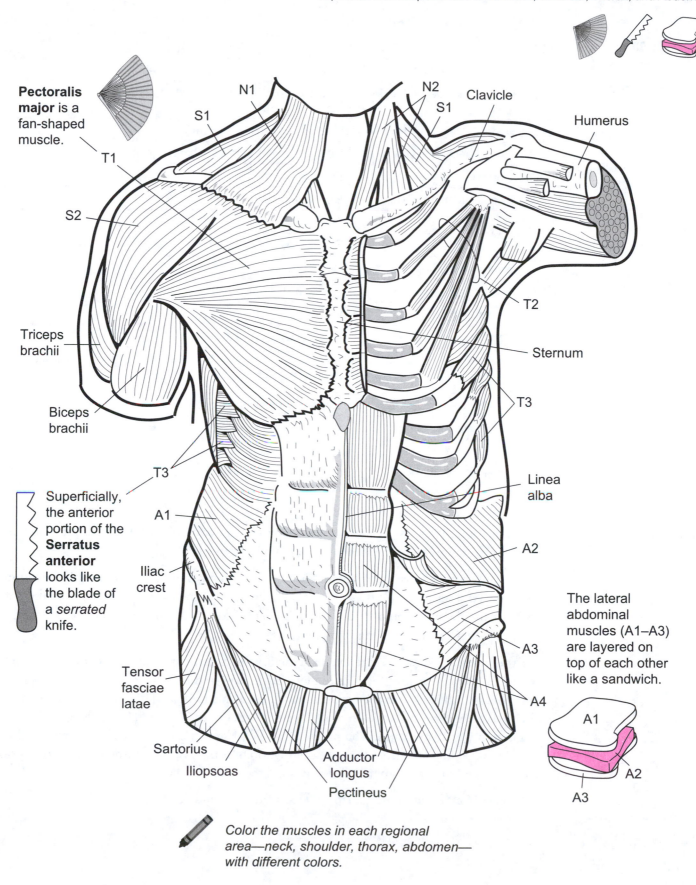

Pectoralis major is a fan-shaped muscle.

N1

N2

S1

S1

Clavicle

Humerus

T1

S2

Triceps brachii

Biceps brachii

T2

Sternum

T3

T3

Linea alba

Superficially, the anterior portion of the **Serratus anterior** looks like the blade of a *serrated* knife.

A1

A2

Iliac crest

The lateral abdominal muscles (A1–A3) are layered on top of each other like a sandwich.

A3

Tensor fasciae latae

A4

A1

Sartorius

A2

Iliopsoas

Adductor longus

A3

Pectineus

Color the muscles in each regional area—neck, shoulder, thorax, abdomen—with different colors.

Description

Muscle Name	Action
1. Coracobrachialis	Flexion and abduction of arm/shoulder
2a. Biceps brachii, short head	Flexes and supinates forearm *(turns the corkscrew and pulls out the cork)*
2b. Biceps brachii, long head	Flexes and supinates forearm
3a. Triceps brachii, long head	Extends forearm; stabilizes shoulder joint
3b. Triceps brachii, medial head	Extends forearm
3c. Triceps brachii, lateral head	Extends forearm
4. Brachialis	Flexes forearm
5. Pronator teres	Flexes forearm; pronates forearm
6. Brachioradialis	Flexes forearm
7. Flexor carpi radialis	Flexes wrist; abducts hand
8. Palmaris longus	Weak wrist flexor
9. Flexor carpi ulnaris	Flexes and adducts hand
10. Anconeus	Adducts ulna during forearm rotation; weak forearm extensor
11. Brachioradialis	Flexes forearm
12. Extensor carpi radialis longus	Extends wrist; abducts hand
13. Flexor carpi ulnaris	Flexes and abducts hand
14. Extensor carpi ulnaris	Extends and adducts hand
15. Extensor digitorum	Extends hand, extends digits 2–5

Study Tips

- Most muscles that act as **flexors** are best seen in the anterior view of the upper limb, and muscles that act as **extensors** are best seen on the posterior view of the upper limb.

- In the anterior view, use the **brachioradialis** as a landmark in the forearm. It is the widest muscle on the lateral surface and inserts on the styloid process of the radius. Medially from the brachioradialis, use the mnemonic "**F**oolish **P**eople **F**ollow" for the following muscles: **F**lexor carpi radialis, **P**almaris longus, and **F**lexor carpi ulnaris.

- In the posterior view, use the **extensor digitorum** muscle as a landmark for the forearm. To correctly identify it, find the four tendons anchoring to the phalanges in all the fingers except the thumb. All these tendons are associated with this muscle. Then, use the mnemonic "**E**at **F**ood, **E**xcept **B**roccoli" to learn the adjacent muscles in the forearm. Laterally from the extensor digitorum is the first part of the phrase, "**E**at **F**ood," for the **E**xtensor carpi ulnaris and **F**lexor carpi ulnaris. Medially from the extensor digitorum is the last part of the phrase, "**E**xcept **B**roccoli" for the **E**xtensor carpi radialis longus and the **B**rachioradialis.

- The **PALM**aris *longus* muscle anchors directly into the middle of the PALM.

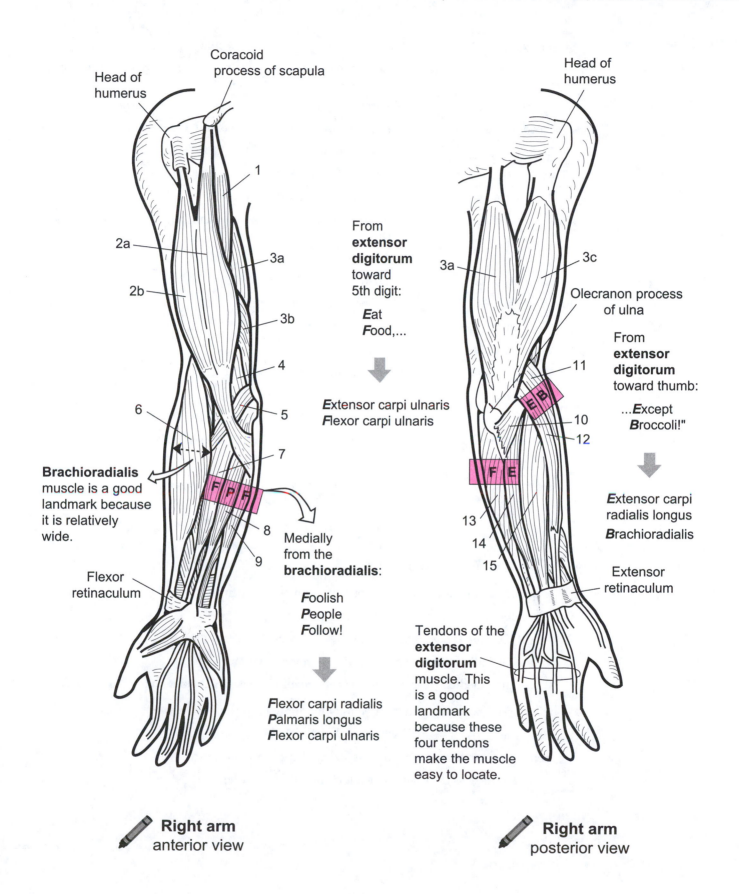

Head of humerus

Coracoid process of scapula

1

2a

2b

3a

3b

4

6

5

7

Brachioradialis muscle is a good landmark because it is relatively wide.

8

9

Flexor retinaculum

From **extensor digitorum** toward 5th digit:

*E*at *F*ood,...

⬇

*E*xtensor carpi ulnaris
*F*lexor carpi ulnaris

Medially from the **brachioradialis**:

*F*oolish
*P*eople
*F*ollow!

⬇

*F*lexor carpi radialis
*P*almaris longus
*F*lexor carpi ulnaris

✏ **Right arm**
anterior view

Head of humerus

3a

3c

Olecranon process of ulna

From **extensor digitorum** toward thumb:

...*E*xcept *B*roccoli!"

11

10

12

13

14

15

⬇

*E*xtensor carpi radialis longus
*B*rachioradialis

Extensor retinaculum

Tendons of the **extensor digitorum** muscle. This is a good landmark because these four tendons make the muscle easy to locate.

✏ **Right arm**
posterior view

Description

Muscle Name	Action
1. Psoas major	Flexes thigh or flexes trunk on thigh (as in during a bow); also effects lateral flexion of vertebral column
2. Iliacus	Flexes thigh or flexes trunk on thigh (as in during a bow)
3. Tensor fasciae latae	Abducts, flexes, and medially rotates thigh
4. Sartorius	Flexes and laterally rotates thigh, flexes knee
5. Pectineus	Adducts, flexes, and medially rotates thigh
6. Adductor longus	Adducts, flexes, and laterally rotates thigh
7. Adductor magnus	Anterior part flexes and medially rotates thigh; posterior part extends and laterally rotates thigh
8. Gracilis	Adducts hip and flexes leg
9. Rectus femoris	Extends knee and flexes thigh at hip
10. Vastus lateralis	Extends leg at knee
11. Vastus intermedius	Extends leg at knee
12. Vastus medialis	Extends leg at knee
13. Gluteus medius	Abducts and medially rotates hip
14. Gluteus maximus	Extends and laterally rotates hip
15. Gracilis	Adducts hip and flexes leg
16. Semimembranosus	Extends thigh; flexes knee; medially rotates leg
17. Semitendinosus	Extends hip and flexes knee
18. Biceps femoris	Long head = extends hip and flexes knee Short head = flexes knee

Study Tips

- The **sartorius** muscle looks like a **s**ash

- The **quadriceps femoris** is a group of four muscles on the anterior thigh:

    rectus femoris (name indicates location—femoral region)

    vastus **lateralis** (name indicates location—on **lateral** aspect of thigh)

    vastus **medialis** (name indicates location—on **medial** aspect of thigh)

    vastus **intermedius** (located deep and **intermediate** to vastus lateralis and vastus medialis)

- The **hamstrings** are a group of three muscles of the posterior thigh:

    Biceps femoris (is located **by** itself on the lateral aspect of the thigh)

    Semimembranosus (is the **most medial** of all the hamstring muscles)

    Semi**tendin**osus (is the hamstring muscle with the longest **tendon**)

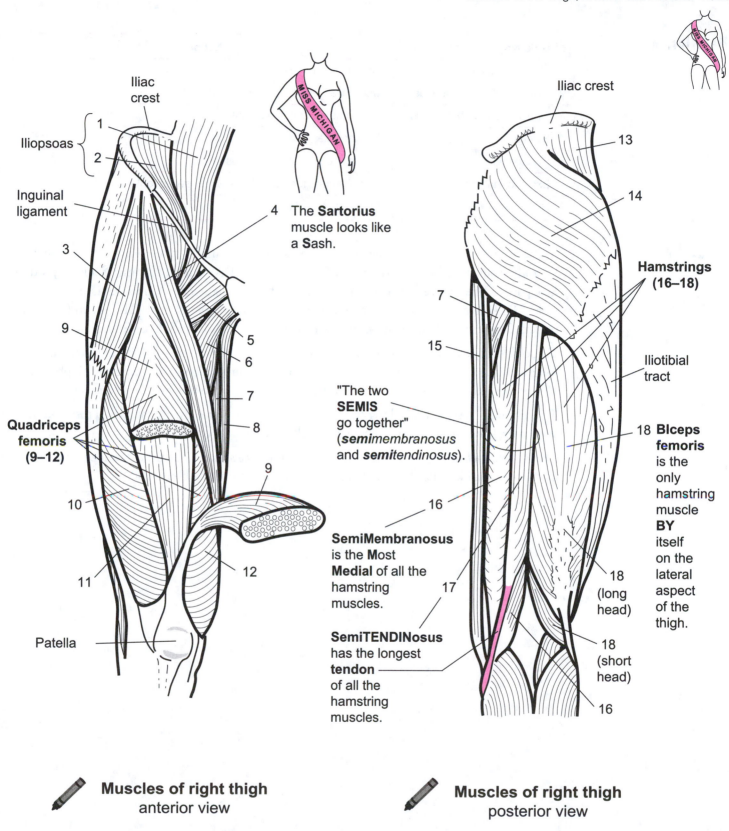

Iliac crest

Iliopsoas { 1 2

Inguinal ligament

3

9

Quadriceps femoris (9–12)

10

11

Patella

4 The **Sartorius** muscle looks like a **S**ash.

5
6
7
8

9

12

Muscles of right thigh
anterior view

Iliac crest

13

14

Hamstrings (16–18)

7

15

Iliotibial tract

"The two **SEMIS** go together" (*semimembranosus* and *semitendinosus*).

SemiMembranosus is the **M**ost **Medial** of all the hamstring muscles.

SemiTENDINosus has the longest **tendon** of all the hamstring muscles.

16

17

18 **BIceps femoris** is the only hamstring muscle **BY** itself on the lateral aspect of the thigh.

18 (long head)

18 (short head)

16

Muscles of right thigh
posterior view

135

Superficial Muscles that Move the Ankle, Foot, and Toes, Anterior and Posterior Views

Description

Muscle Name	Action
1. Tibialis anterior	Dorsiflexes and inverts foot
2. Extensor digitorum longus	Dorsiflexes, everts foot, and extends digits 2–5
3. Fibularis (peroneus) longus	Plantar flexes and everts foot
4. Soleus	Plantar flexes foot
5. Gastrocnemius	Flexes knee and plantar flexes foot
6. Extensor hallicus longus	Extends great toe; dorsiflexes ankle; everts foot
7. Plantaris	Plantar flexion of foot

Study Tips

- In the anterior view, use the **tibia** as a landmark to learn muscles in a sequence either medially or laterally from the tibia. Use the following mnemonic: **T**ake **E**than **F**ishing: **S**ounds **G**ood! The first part of the phrase, **T**ake **E**than **F**ishing, gives the sequence of muscles laterally from the tibia (**T**ibialis anterior, **E**xtensor digitorum longus, **F**ibularis longus). The second part of the phrase, **S**ounds **G**ood, gives the sequence medially from the tibia (**S**oleus, **G**astrocnemius).

- To double-check that you have identified the muscles correctly, use the **extensor digitorum longus** muscle as a landmark. To locate it, find the four tendons anchoring to the phalanges in all the toes except the great toe. Follow these tendons up into this muscle to correctly identify it.

- The **fibularis longus** has a long tendon that inserts into the 5th metatarsal. This tendon *loops around the lateral malleolus* of the fibula, which makes it easy to locate.

- The **soleus** (*soleus* = fish) is so named because it looks like a flat fish. It is located *deep* to the gastrocnemius like a flat fish would rest *deep* on the bottom of a body of water.

Laterally from the tibia:

Take
Ethan
Fishing:
→
Tibialis anterior
Extensor digitorum longus
Fibularis longus

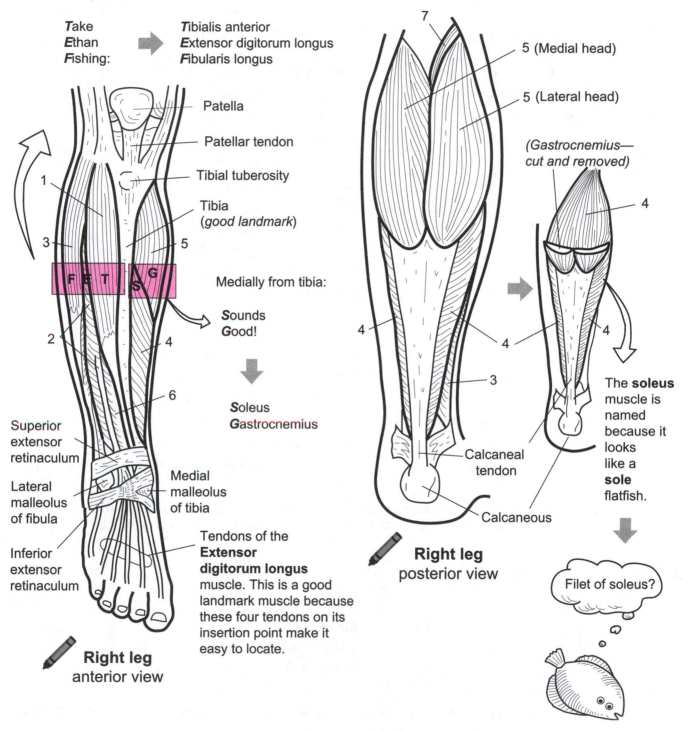

Patella

Patellar tendon

Tibial tuberosity

Tibia
(*good landmark*)

Medially from tibia:

Sounds
Good!

↓

Soleus
Gastrocnemius

Superior
extensor
retinaculum

Lateral
malleolus
of fibula

Inferior
extensor
retinaculum

Medial
malleolus
of tibia

Tendons of the
**Extensor
digitorum longus**
muscle. This is a good
landmark muscle because
these four tendons on its
insertion point make it
easy to locate.

Right leg
anterior view

5 (Medial head)

5 (Lateral head)

*(Gastrocnemius—
cut and removed)*

The **soleus**
muscle is
named
because it
looks
like a
sole flatfish.

Calcaneal
tendon

Calcaneous

Right leg
posterior view

Filet of soleus?

Superficial and Deep Muscles of the Neck, Shoulder, Back, and Gluteal Region

Description

Muscle Name	Action
1. Sternocleidomastoid	Simultaneous contraction of both heads: flexes neck forward. Individual action of each head: rotates head to shoulder on opposite side
2. Trapezius	Elevates, retracts, depresses, or rotates scapula upward; elevates clavicle; extends neck
3. Deltoid	Flexes, extends, abducts; medial and laterally rotates arm
4. Infraspinatus*	Abducts and laterally rotates arm
5. Teres minor*	Adducts, extends, and laterally rotates arm
6. Teres major	Extends, medially rotates, and adducts arm
7. Latissimus dorsi	Extends, adducts, and medially rotates arm
8. External oblique	Compresses anterior abdominal wall; flexes trunk; rotates trunk; depresses lower ribs
9. Gluteus medius	Abducts and medially rotates hip
10. Gluteus maximus	Extends and laterally rotates hip
11. Semispinalis capitis	Extends head and rotates to opposite side
12. Splenius capitis	Extends and hyperextends the head
13. Levator scapulae	Raises scapula and draws it medially; with scapula fixed, flexes neck to same side
14. Rhomboid minor	Retracts, adducts, and stabilizes the scapula
15. Rhomboid major	Adducts, retracts, elevates, and rotates scapula; stabilizes scapula
16. Supraspinatus*	Abducts arm; stabilizes shoulder joint
17. Serratus anterior	Abducts and stabilizes scapula
18. Serratus posterior inferior	Depresses last four ribs
19. Internal oblique	Compresses anterior abdominal wall; flexes trunk; rotates trunk; depresses lower ribs

Rotator cuff muscles: **supraspinatus**, **infraspinatus**, **teres minor**, and **subscapularis**. The only one of these muscles not shown in either the table or the illustration is the **subscapularis**.

Study Tips

- The rotator cuff = the "SITS" muscle group.

S	**S**upraspinatus
I	**I**nfraspinatus
T	**T**eres MINOR
S	**S**ubscapularis

A pro baseball pitcher injured his rotator cuff, so now he **SITS** down in the **MINOR** leagues (*minor* indicates teres *minor* instead of teres *major*)

Superficial and Deep Muscles of the Neck, Shoulder, Back, and Gluteal Region

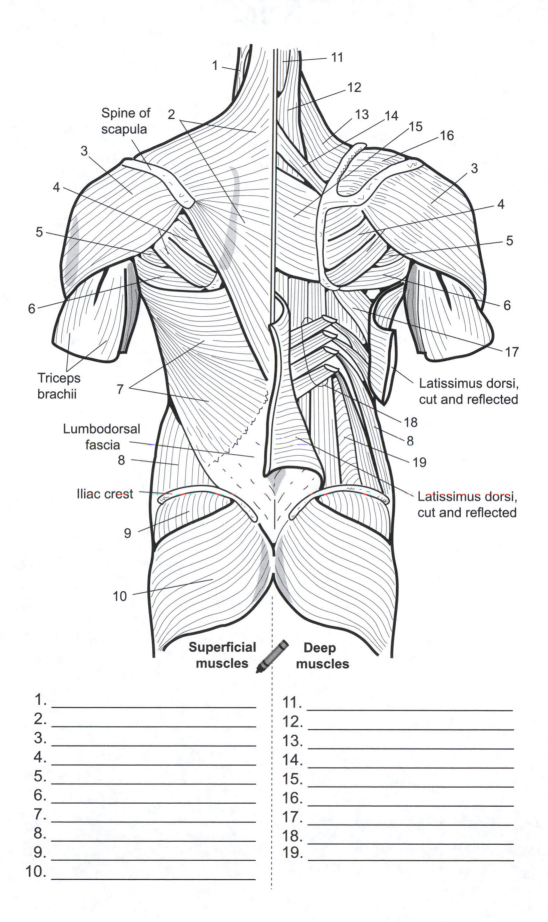

Spine of scapula

Triceps brachii

Lumbodorsal fascia

Iliac crest

Latissimus dorsi, cut and reflected

Latissimus dorsi, cut and reflected

Superficial muscles   **Deep muscles**

1. _____   11. _____
2. _____   12. _____
3. _____   13. _____
4. _____   14. _____
5. _____   15. _____
6. _____   16. _____
7. _____   17. _____
8. _____   18. _____
9. _____   19. _____
10. _____

Notes

Nervous System

Description

There are many different structural types of neurons. For the sake of comparison, we will examine only three different types: **unipolar**, **bipolar**, and **multipolar** neurons. They are named after the number of long cellular processes (*dendrite, axon*) that branch off the cell body. A unipolar neuron has one process that branches into a dendrite and an axon. A bipolar neuron has two processes— one dendrite, one axon. A multipolar neuron has many dendrites and one axon.

Every neuron has three basic parts: a **dendrite**(s), a **cell body**, and an **axon.** A neuron may have more than one dendrite but only one axon. At the end of the axon is the **synaptic knob.** Here there is a small space called a **synapse** that connects the neuron to a muscle or a gland or another neuron.

Nerve impulses travel in a one-way direction. The dendrite is a sensory receptor that receives various types of stimuli. It is where the nervous impulse is generated. From here, the impulse travels into the cell body, then along the axon to the end of the synaptic knob. The synaptic knob produces and releases a chemical called a **neurotransmitter.** This chemical messenger diffuses across the synapse and binds to a receptor in the muscle cell, glandular cell, or neuron to which it is connected. Once the neurotransmitter binds to its receptor it induces a response. For example, it may stimulate muscle to contract or cause cells in a gland to release a hormone or stimulate a neuron to fire and generate a nervous impulse.

Location

- **Unipolar neurons**—*ex:* sensory neurons of the peripheral nervous system

- **Bipolar neurons**—*ex:* photoreceptors in the retina of the eye

- **Multipolar neurons**—*ex:* most common type of neuron in the brain and spinal cord; motor neurons in the peripheral nervous system

Function

All neurons conduct nerve impulses.

Key to Illustration

1. Unipolar neuron   D = Dendrite
2. Bipolar neuron   C = Cell body
3. Multipolar neuron   A = Axon
                          SK = Synaptic knobs

Color all the dendrites one color and the synaptic knobs a different color.

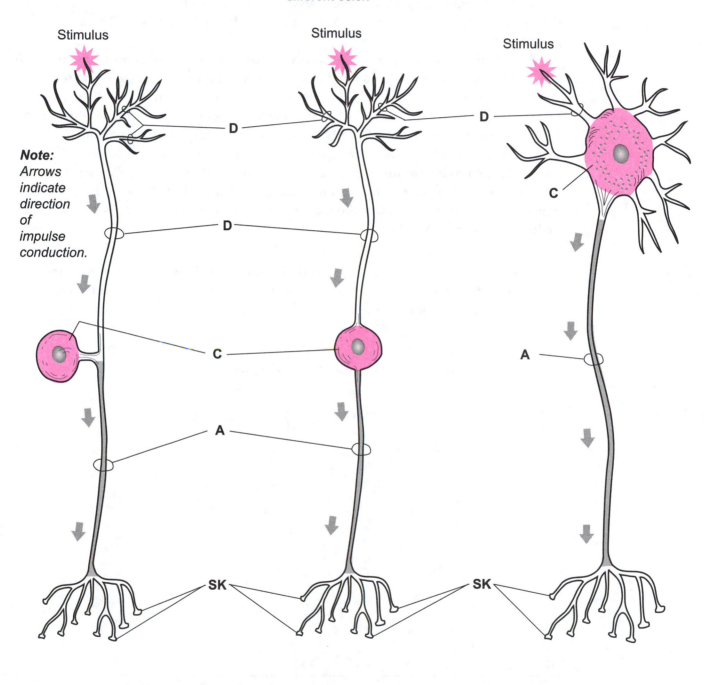

Stimulus

Stimulus

Stimulus

Note: Arrows indicate direction of impulse conduction.

D

D

C

C

A

SK

SK

1. _____

2. _____

3. _____

D = _____
C = _____
A = _____
SK = _____

Description

A multipolar neuron is one type of **neuron**. It has many processes called **dendrites**, which respond to stimuli. The **soma** (*cell body*) is the regional area that includes the nucleus, cytoplasm, and various organelles. The rough endoplasmic reticulum is found in large clusters called **Nissl bodies**. The cell body tapers off into a funnel-shaped structure called the **axon hillock**, which becomes the **axon**. Nervous impulses are conducted along the length of the axon. The axon ends in the **synaptic knob**.

Axons can be either **myelinated** or **unmyelinated**. In the process of **myelination**, the axon is wrapped in a sheath of lipid and protein that insulates the nervous impulse and speeds impulse conduction. This process begins in the fetus and continues into late adolescence. In the central nervous system (CNS) cells called oligodendrocytes are responsible for the myelination process. In the peripheral nervous system (PNS), **neurolemmocytes** (*Schwann cells*) perform this task by wrapping themselves around the axon many times. These layers of plasma (*cell*) membrane constitute the **myelin sheath**. During this process, the nucleus and other organelles are pushed to the outer surface of the neurolemmocyte (*Schwann cell*). This outer layer is called the **neurilemma**. Segments of unwrapped axon between neurolemmocytes (*Schwann cells*) are called **nodes of Ranvier**.

Analogy

Each neuron has only one **axon**. The axon is like an **electrical cord**. The axon conducts a nervous impulse like the copper wires in the cord conduct electricity. The **myelin sheath** serves to insulate the axon like the **plastic casing** around the electrical cord.

Location

Multipolar neurons are the most common type of neuron in the brain and spinal cord but also occur as motor neurons in the peripheral nervous system.

Function

All neurons conduct nerve impulses.

Key to Illustration

1. Dendrite
2. Soma (*cell body*)
3. Nucleus of neuron
4. Nissl bodies
5. Axon hillock
6. Axon
7. Neurolemmocyte (*Schwann cell*)
8. Nodes of Ranvier
9. Nucleus of neurolemmocyte (*Schwann cell*)
10. Neurilemma
11. Endoneurium
12. Synaptic knobs

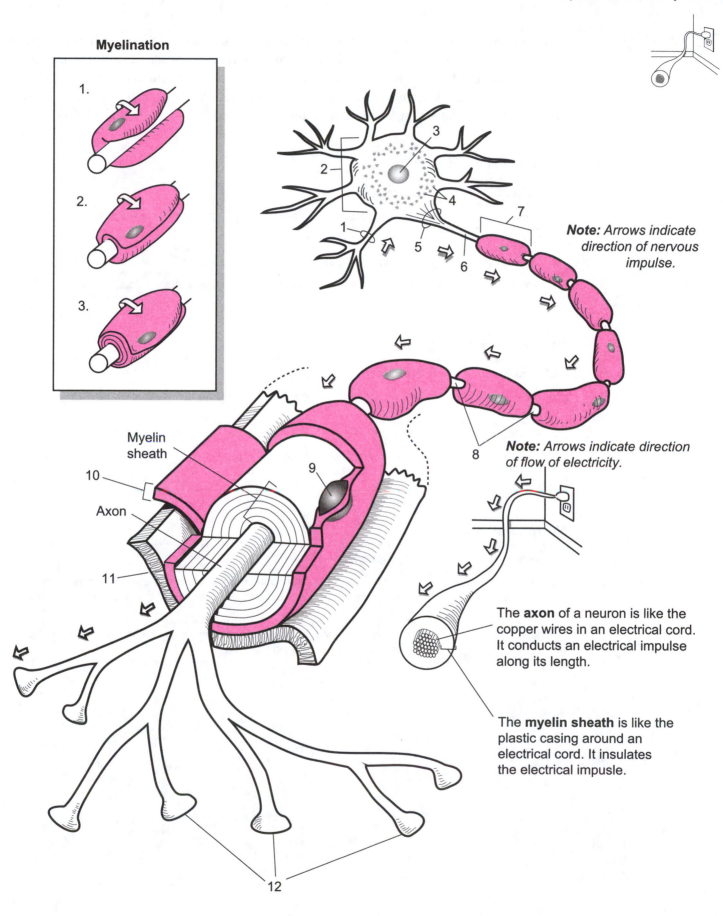

Myelination

1.

2.

3.

Note: Arrows indicate direction of nervous impulse.

Note: Arrows indicate direction of flow of electricity.

Myelin sheath

Axon

The **axon** of a neuron is like the copper wires in an electrical cord. It conducts an electrical impulse along its length.

The **myelin sheath** is like the plastic casing around an electrical cord. It insulates the electrical impusle.

Description

Peripheral nerves spread throughout the body like electrical wiring through a house. They are a collection of bundles of microscopic axons. Examples include the sciatic and femoral nerves in the thigh and the brachial nerve in the arm. Nerves are visible to the naked eye and structured as a series of tubes within tubes. Each tube-like structure is wrapped in a protective connective tissue. Each of the labeled structures in the illustration is explained in the tables below.

Tubes within tubes

Structure	Description
1. Nerve	long, macroscopic, cable-like structure containing bundles of axons
2. Fascicle	a single bundle of axons
3. Myelinated axon	an axon wrapped in a protective myelin sheath
4. Axon	long, thin part of a neuron that carries nerve impulses along its length

Connective tissue organization

Structure	Description
5. Epineurium	the thick layer of dense irregular connective tissue that wraps around the outside of a nerve
6. Perineurium	the cellular connective tissue layer that wraps around each fascicle
7. Endoneurium	the thin layer of areolar connective tissue that wraps around each axon and binds one myelinated axon to another within a fascicle

Neuronal structures

Structure	Description
8. Neurolemmocyte (Schwann cell)	a cell that wraps itself around the axon of a neuron numerous times during development of the nervous system. The end result is that a myelin sheath is created
9. Myelin sheath	the layers of the neurolemmocyte's cell membrane that form a protective, insulting coating of lipoprotein around the axon

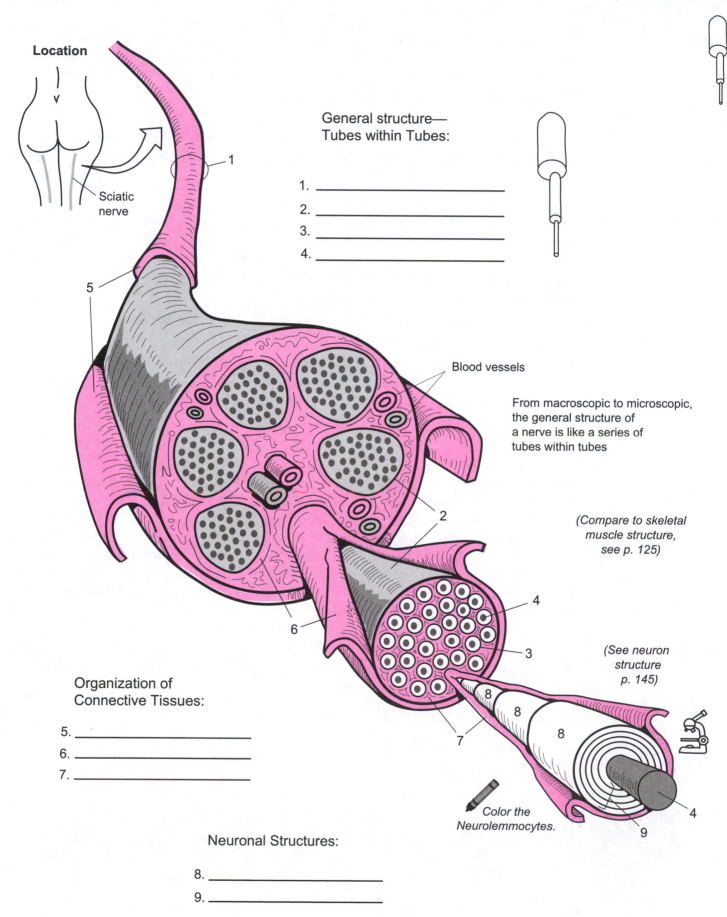

Location

Sciatic nerve

General structure—
Tubes within Tubes:

1. _____
2. _____
3. _____
4. _____

Blood vessels

From macroscopic to microscopic,
the general structure of
a nerve is like a series of
tubes within tubes

*(Compare to skeletal
muscle structure,
see p. 125)*

*(See neuron
structure
p. 145)*

Organization of
Connective Tissues:

5. _____
6. _____
7. _____

Neuronal Structures:

8. _____
9. _____

*Color the
Neurolemmocytes.*

Description

The **spinal cord** is a long, slender structure that links the body and the brain. Most of the cord is protected by the bony vertebrae because it runs through the vertebral canal of the vertebral column. Three layers of protective membranes called **meninges** surround the spinal cord and brain. From outermost to innermost, these are as follows:

- **Dura mater:** thickest and strongest; contains fibrous connective tissue
- **Arachnoid:** thin layer made of simple squamous epithelium; lacks blood vessels
- **Pia mater:** tightly adheres to the spinal cord and follows every surface feature; supplies many blood vessels directly to the spinal cord.

Below the arachnoid is a potential space called the **subarachnoid space**, which is filled with **cerebrospinal fluid**. This serves as a cushion to protect the spinal cord and functions as a medium through which to deliver nutrients and remove wastes. Extending laterally off the spinal cord are 31 pairs of **spinal nerves**. These become the various peripheral nerves that spread throughout the body. The spinal nerves and their associated structures are:

- **Dorsal root:** contains only *sensory* axons.
- **Dorsal root ganglion:** contains the neuron cell bodies of sensory neurons.
- **Ventral root:** contains only motor axons.
- **Dorsal ramus:** branches off the spinal nerves that innervate muscles and skin of the back.
- **Rami communicantes:** branch off the spinal nerves that contain axons related to the autonomic nervous system (ANS).

The spinal cord contains areas of **gray matter** and **white matter**. The white matter is located in the outer portion of the spinal cord and consists of myelinated axons that run along its length. It is divided into the following three regional areas called **funiculi** (sing. *funiculus*): a **posterior funiculus**, **lateral funiculi**, and an **anterior funiculus**. The **white commissure** is a narrow band of white matter that connects the anterior funiculi together.

The **gray matter** is located in the inner portion of the spinal cord and includes short neurons called interneurons along with cell bodies, dendrites, and axon terminals of other neurons. The three regional areas of gray matter are referred to as horns: **posterior horn**, **lateral horn**, **anterior horn**. In the center of the spinal cord is a small passageway called the **central canal**, which contains cerebrospinal fluid. The horizontal band of gray matter that surrounds the central canal is called the **gray commissure**.

Analogy

The gray matter in the middle of the spinal cord looks like a butterfly. Note that the exact shape of the gray matter changes from one segment of the spinal cord to another so it does not always look exactly like a butterfly.

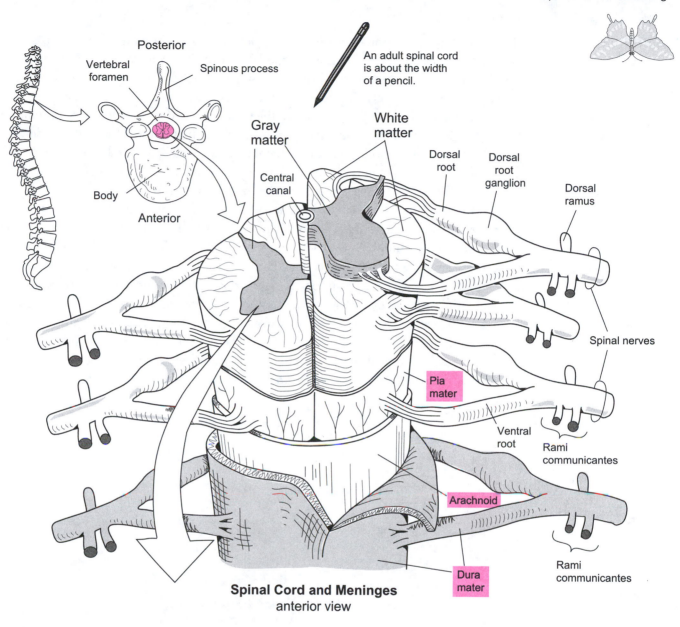

Posterior

Vertebral foramen

Spinous process

Body

Anterior

An adult spinal cord is about the width of a pencil.

Gray matter

White matter

Central canal

Dorsal root

Dorsal root ganglion

Dorsal ramus

Spinal nerves

Pia mater

Ventral root

Rami communicantes

Arachnoid

Dura mater

Rami communicantes

Spinal Cord and Meninges
anterior view

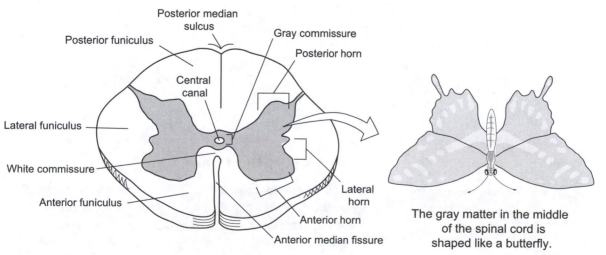

Posterior median sulcus

Gray commissure

Posterior funiculus

Posterior horn

Central canal

Lateral funiculus

Lateral horn

White commissure

Anterior funiculus

Anterior horn

Anterior median fissure

Spinal Cord
cross-section

The gray matter in the middle of the spinal cord is shaped like a butterfly.

Description

The illustration shows three of the major regional areas of the brain: *brain stem*, *cerebellum*, and *cerebrum*. The **brain stem** is located at the base of the brain and contains regulatory centers to control things we take for granted such as respiration, cardiovascular activities, and digestion. It also relays information between the cerebellum and the cerebrum.

The **cerebellum** is located posterior to the brain stem and inferior to the cerebrum. It is divided into two left and right halves, or **hemispheres**, and is highly folded to increase surface area. Its general function is to work with the cerebrum to coordinate skeletal muscle movements but it also allows the body to maintain proper balance and posture.

The **cerebrum** is the largest part of the brain and contains billions of neurons. Like the cerebellum, it is divided into two halves, or **hemispheres**. The deep division between these two hemispheres is called the **longitudinal fissure**. The term **fissure** is used to indicate a very deep groove or depression that separates major sections of the brain. The surface of the cerebrum is not smooth; it is folded into many little hills and gulleys. Each hill is called a **convolution**, or **gyrus**, and each gulley is a shallow groove called a **sulcus**. The cerebrum is the part of the brain associated with higher brain functions, such as planning, reasoning, analyzing, and storing/accessing memories.

Ironically, without the cerebrum, you would not be able to read and learn about the brain as you are doing right now. It also receives and interprets sensory information and coordinates various motor functions, such as those involved in speech. The cerebrum is divided into four major lobes named after the bones that cover them: *frontal*, *parietal*, *temporal*, and *occipital*.

Key to Illustration

Surface features (SF)	Major brain regions (R)	Lobes of cerebrum (L)
SF1. Gyrus *(convolution)*	R1. Cerebrum	L1. Frontal lobes
SF2. Sulcus	R2. Brain stem	L2. Parietal lobes
	R3. Cerebellum	L3. Temporal lobes
		L4. Occipital lobes

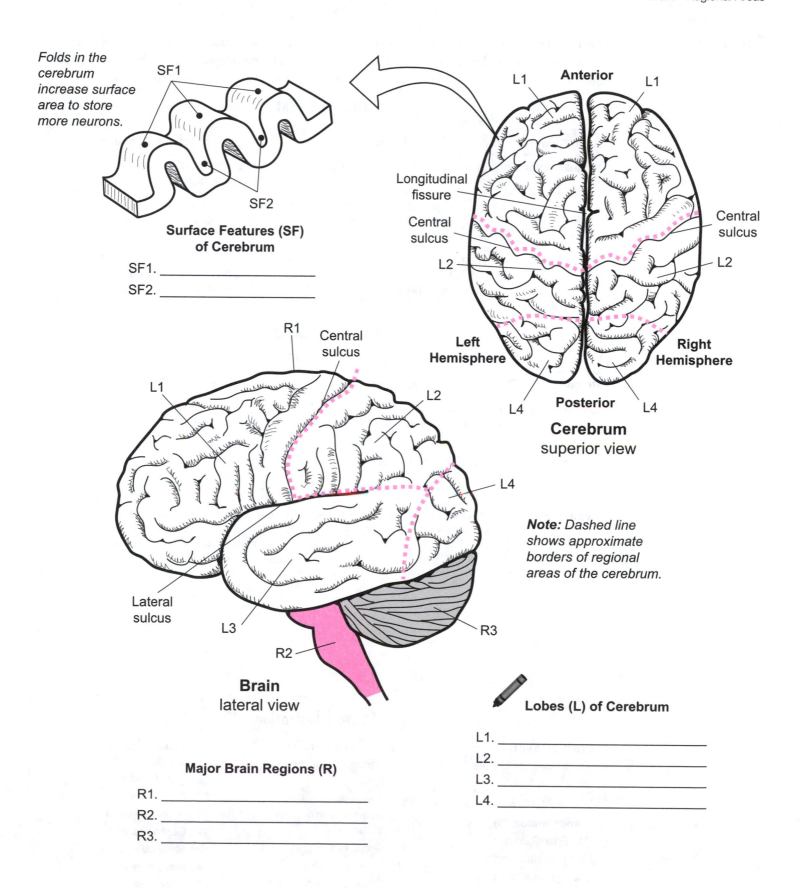

Folds in the cerebrum increase surface area to store more neurons.

SF1

SF2

Surface Features (SF) of Cerebrum

SF1. _____

SF2. _____

L1   **Anterior**   L1

Longitudinal fissure

Central sulcus

Central sulcus

L2

L2

Left Hemisphere

Right Hemisphere

L4   **Posterior**   L4

Cerebrum
superior view

R1

Central sulcus

L1

L2

L4

Note: *Dashed line shows approximate borders of regional areas of the cerebrum.*

Lateral sulcus

L3

R2

R3

Brain
lateral view

Lobes (L) of Cerebrum

L1. _____

L2. _____

L3. _____

L4. _____

Major Brain Regions (R)

R1. _____

R2. _____

R3. _____

Description

The midsagittal section of the brain reveals many different structures. It is especially good for viewing the structural relationships between the **brain stem, diencephalon, cerebrum,** and **cerebellum.** The tables below summarize key information about the brain stem and diencephalon.

BRAIN STEM

Brain Stem Region	Description	General Functions
Medulla oblongata	Between spinal cord and pons	Respiratory control center; cardiovascular control center
Pons	Between medulla and midbrain; bulges out as widest region in brain stem	Controls respiration along with medulla; relays information from cerebrum to cerebellum
Midbrain	Between diencephalon and pons; includes corpora quadrigemina and cerebral aqueduct	Visual and auditory reflex centers; provides pathway between brain stem and cerebrum

DIENCEPHALON

Diencephalon Region	Description	General Functions
Epithalamus	Roof of third ventricle; includes pineal gland; choroid plexus found here (*forms cerebrospinal fluid*)	Pineal gland makes hormone melatonin, which regulates day-night cycles;
Thalamus	Two egg-shaped bodies that surround the third ventricle	Relays sensory information to cerebral cortex; relays information for motor activities; information filter
Hypothalamus	Forms floor of third ventricle; between thalamus and optic chiasm	Controls autonomic centers for heart rate, blood pressure, respiration, digestion; hunger center; thirst center; regulation of body temperature; production of emotions

Key to Illustration

Brain Stem (B)
B1. Medulla oblongata
B2. Pons
B3. Midbrain

Diencephalon (D)
D1. Epithalamus
D2. Thalamus
D3. Hypothalamus

1. Anterior commissure
2. Septum pellucidum
3. Fornix
4. Interthalamic adhesion (*intermediate mass of thalamus*)
5. Corpus callosum
6. Pineal gland (*body*)
7. Corpora quadrigemina

8. Cerebral aqueduct (*aqueduct of Sylvius*)
9. Transverse fissure
10. Arbor vitae
11. Fourth ventricle
12. Central canal
13. Mammillary body
14. Pituitary gland
15. Optic chiasm

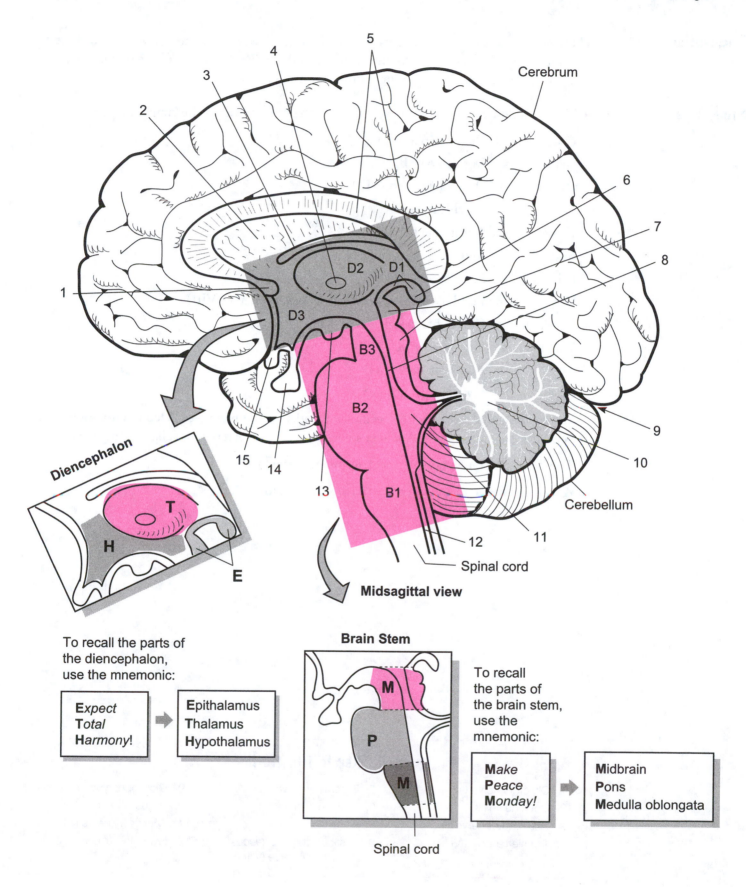

Cerebrum

Cerebellum

Spinal cord

Midsagittal view

Diencephalon

To recall the parts of
the diencephalon,
use the mnemonic:

Expect	→	**E**pithalamus
Total		**T**halamus
Harmony!		**H**ypothalamus

Brain Stem

Spinal cord

To recall
the parts of
the brain stem,
use the
mnemonic:

Make	→	**M**idbrain
Peace		**P**ons
Monday!		**M**edulla oblongata

Description

There are 12 pairs of **cranial nerves** that are best observed on the inferior view of a whole brain. Beginning near the frontal lobe of the cerebrum and moving down toward the spinal cord, they are numbered using Roman numerals from one (I) to twelve (XII).

Study Tips

- Use the following mnemonic device to recall the proper order of the cranial nerves:

Oscar's	=	**O**lfactory nerve (I)
Old	=	**O**ptic nerve (II)
Ostrich	=	**O**culomotor nerve (III)
Tasted	=	**T**rochlear nerve (IV)
Tomatoes	=	**T**rigeminal nerve (V)
And	=	**A**bducens nerve (VI)
Felt	=	**F**acial nerve (VII)
Very	=	**V**estibulocochlear (*acoustic* or *auditory*) nerve (VIII)
Good,	=	**G**lossopharyngeal nerve (IX)
Vomited	=	**V**agus nerve (X)
Any	=	**A**ccessory nerve (XI)
How	=	**H**ypoglossal nerve (XII)

- Associate cranial nerves with specific landmarks on the brain—*ex:* **Oculomotor nerve** (III) is below the mamillary body, **Abducens nerve** (VI) is between the medulla and the pons

- The **T**hickest cranial nerve is the **Trigeminal nerve** (V)

- **Accessory nerve** (XI) runs parallel to the spinal cord

Key to Illustration

1. Olfactory nerve (I)
2. Optic nerve (II)
3. Oculomotor nerve (III)
4. Trochlear nerve (IV)
5. Trigeminal nerve (V)
6. Abducens nerve (VI)
7. Facial nerve (VII)
8. Vestibulocochlear (*acoustic* or *auditory)* nerve (VIII)
9. Glossopharyngeal nerve (IX)
10. Vagus nerve (X)
11. Accessory nerve (XI)
12. Hypoglossal nerve (XII)

To recall the cranial nerves from the first pair to the last pair, use the following mnemonic:

*O*scar's
*O*ld
*O*strich
*T*asted
*T*omatoes
*A*nd
*F*elt
*V*ery
*G*ood,
*V*omited
*A*ny
*H*ow

Color the matching pairs of cranial nerves.

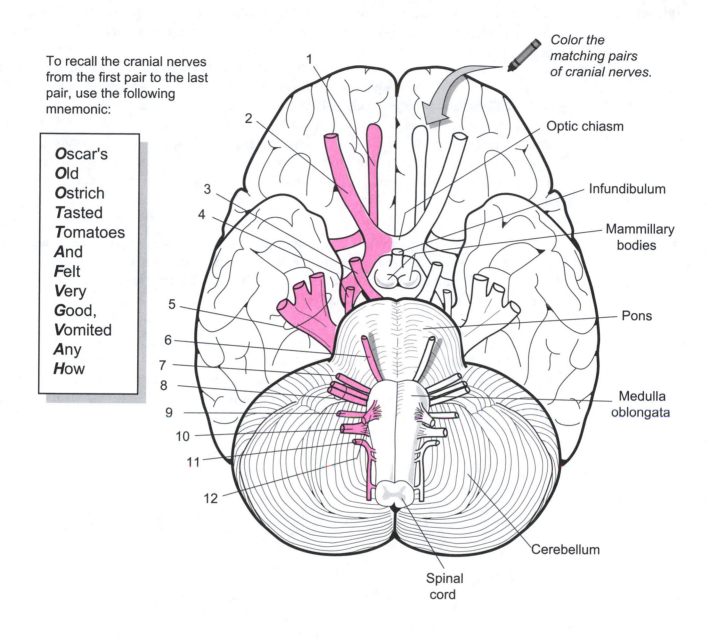

Optic chiasm

Infundibulum

Mammillary bodies

Pons

Medulla oblongata

Cerebellum

Spinal cord

Cranial Nerves

1. _____
2. _____
3. _____
4. _____
5. _____
6. _____
7. _____
8. _____
9. _____
10. _____
11. _____
12. _____

Description

The heart contains ventricles that fill with blood, while the brain contains ventricles that are constantly filled with cerebrospinal fluid. In total, the brain has four ventricles inside it: *lateral ventricle* (of left hemisphere), *lateral ventricle* (of right hemisphere), *third ventricle*, and *fourth ventricle*. This entire network is referred to as the **ventricular system** in the brain. The lateral ventricles are the largest of the four and do not directly connect to each other as they are separated by a thin partition called the **septum pellucidum**. Both do connect to the third ventricle in the region of the diencephalon by small passageways called **interventricular foramina**. The third ventricle is connected to the fourth ventricle by a passageway called the **cerebral aqueduct** (*aqueduct of Sylvius*). The fourth ventricle is located in the pons (of the brain stem) and the cerebellum. It communicates with a very narrow passageway called the **central canal**, which runs through the middle of the spinal cord.

Analogy

To visualize the relative positions of the ventricles, compare the whole **ventricular system** to the **hollow head of a ram**. The **fourth ventricle** is like the **neck of the ram**, the **third ventricle** is like the head, and the **lateral ventricles** are like the **two horns**. The ram's horns also follow the same general shape of the paired lateral ventricles.

Study Tip

The first and second ventricles are not numbered because they are the lateral ventricles. If you think of the two lateral ventricles as *first ventricle* and *second ventricle*, the numbering makes sense in relation to the **third ventricle** and **fourth ventricle**. Ah, the goofy things that anatomists do! As the saying goes, "you are not a good anatomist unless you know 87 different names for the same structure."

Key to Illustration

1. Lateral ventricles
1a. Anterior horns of lateral ventricles
1b. Posterior horns of lateral ventricles
1c. Inferior horns of lateral ventricles

2. Third ventricle
3. Cerebral aqueduct (*aqueduct of Sylvius*)
4. Fourth ventricle
5. Central canal

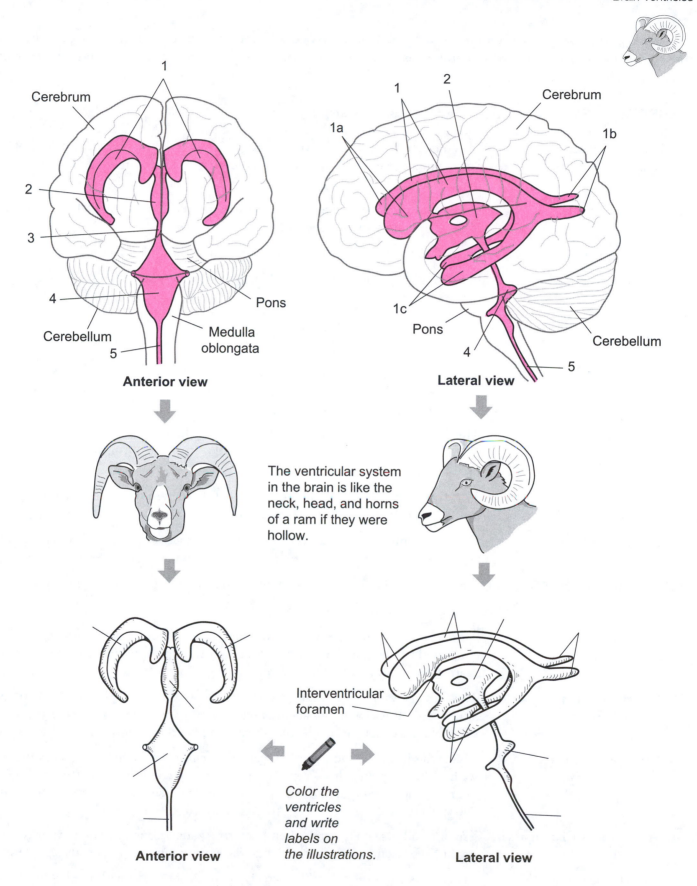

Cerebrum

1

2

3

4

Cerebellum

5

Pons

Medulla
oblongata

Anterior view

1a

1

2

Cerebrum

1b

1c

Pons

4

5

Cerebellum

Lateral view

The ventricular system in the brain is like the neck, head, and horns of a ram if they were hollow.

Interventricular foramen

Color the ventricles and write labels on the illustrations.

Anterior view

Lateral view

This module will describe some of the selected functional areas of the cerebral cortex. These areas have been divided into three general groups: **sensory areas**, **motor areas**, and **association areas**. Note that the words *cortex* and *area* are often used interchangeably.

SENSORY AREAS   Control regions where sensations are perceived

1. **Primary somatic sensory cortex**	This important region is shown in dark gray behind the central sulcus. General sensory input (e.g. touch, temperature, pressure, and pain) from all parts of the body is perceived here.
2. **Gustatory cortex**	Located in the parietal lobe; taste sensations are perceived here, such as the flavors of the ice cream shown in the icon
3. **Auditory cortex**	Located in the temporal lobe; auditory stimuli are processed by the brain here
4. **Visual cortex**	Located in the occipital lobe; visual images are perceived here (like the star shown in the icon)

MOTOR AREAS   Control centers for conscious muscle movements

1. **Primary motor cortex**	This important area is shown in color in front of the central sulcus. It controls voluntary muscle movements throughout the body, including those of the hands and feet, arms and legs, face and tongue.
2. **Premotor cortex**	This area serves as the "choreographer" for the primary motor cortex. It decides which muscle groups will be used and how they will be used prior to stimulating the primary motor cortex.
3. **Motor speech area** (*Broca's area*)	This area controls and coordinates the muscles involved in normal, fluent speech. Damage to this area can result in strained speech with disconnected words.
4. **Frontal eye field**	This area controls muscle movements of the eye, such as those needed to read this page

ASSOCIATION AREAS   Control regions—near sensory areas—involved in recognizing and analyzing incoming information

1. **Prefrontal area**	This area is most highly developed in humans and other primates. It regulates emotional behavior and mood and also is involved in planning, learning, reasoning, motivation, personality, and intellect.
2. **Somatic sensory association area**	This area allows you to *predict* that sandpaper is rough, for example, even without looking at it. It also stores memories about previous sensory experiences so you can determine when blindfolded, for example, that the object placed in your hand was a pair of scissors.
3. **Sensory speech area** (*Wernicke's area*)	This area seems to be an important part of language development—processing words we hear being spoken. It also appears important for children when they are sounding out new words. Damage to this area may result in deficiencies in recognizing written and spoken words.
4. **Auditory association area**	This area allows you to comprehend, interpret, analyze, and question what you are hearing. For example, it enables you to recognize a familiar song or disregard noise.
5. **Visual association area**	This allows you to associate the perceived image of the star with the letters "S-T-A-R". You connect the word "star" with the image of a star.

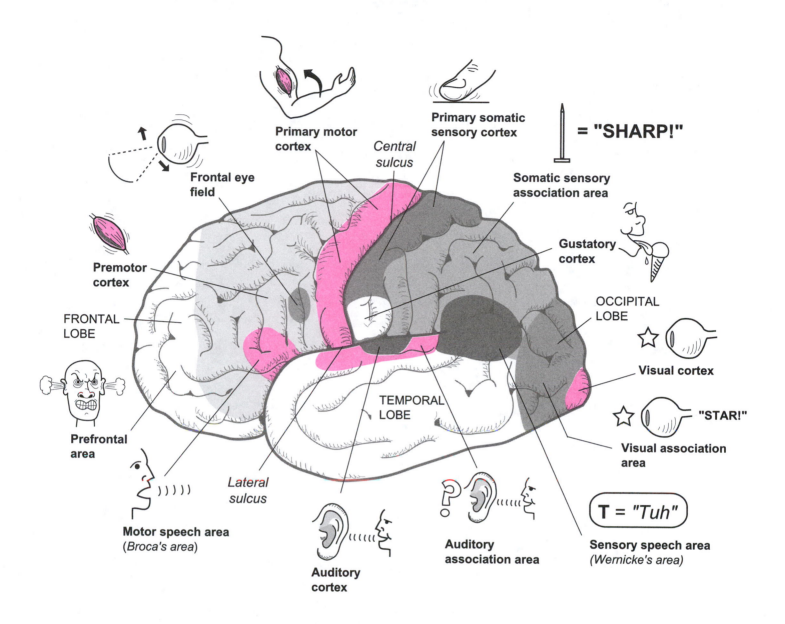

Left Cerebral Hemisphere
lateral view

Primary motor cortex

Central sulcus

Primary somatic sensory cortex

= "SHARP!"

Frontal eye field

Somatic sensory association area

Gustatory cortex

Premotor cortex

OCCIPITAL LOBE

FRONTAL LOBE

Visual cortex

Prefrontal area

"STAR!"

Visual association area

TEMPORAL LOBE

Lateral sulcus

T = *"Tuh"*

Motor speech area (*Broca's area*)

Auditory cortex

Auditory association area

Sensory speech area (*Wernicke's area*)

159

Notes

Endocrine System

Description   Along with the nervous system, the endocrine system is one of the great regulators of activities in the human body. Its major purpose is to maintain homeostasis and to regulate various processes such as growth and human development. It consists of many different glands containing specific cells that synthesize and release chemical messengers called hormones. These hormones enter the bloodstream, where they travel to a specific cell called a target cell. This target cell has a receptor for that specific hormone. Once the hormone binds to the receptor, it induces a response in that cell. The general concept of how the endocrine system functions is illustrated below:

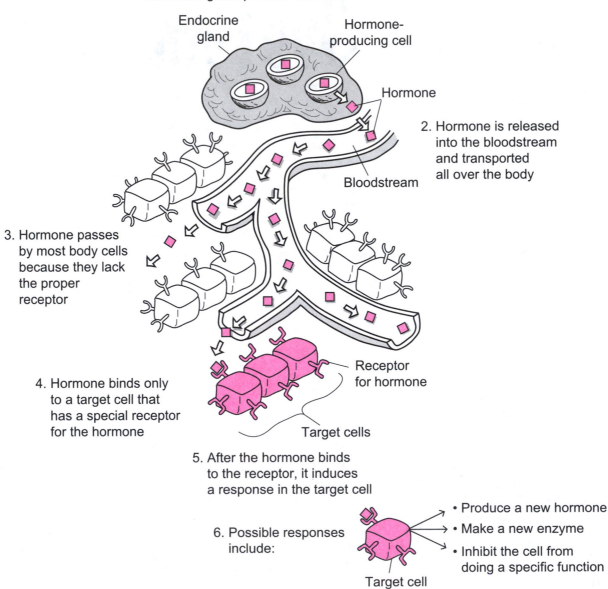

1. Hormone-producing cells within an endocrine gland produce a hormone

Endocrine gland

Hormone-producing cell

Hormone

2. Hormone is released into the bloodstream and transported all over the body

Bloodstream

3. Hormone passes by most body cells because they lack the proper receptor

4. Hormone binds only to a target cell that has a special receptor for the hormone

Receptor for hormone

Target cells

5. After the hormone binds to the receptor, it induces a response in the target cell

6. Possible responses include:

• Produce a new hormone
• Make a new enzyme
• Inhibit the cell from doing a specific function

Target cell

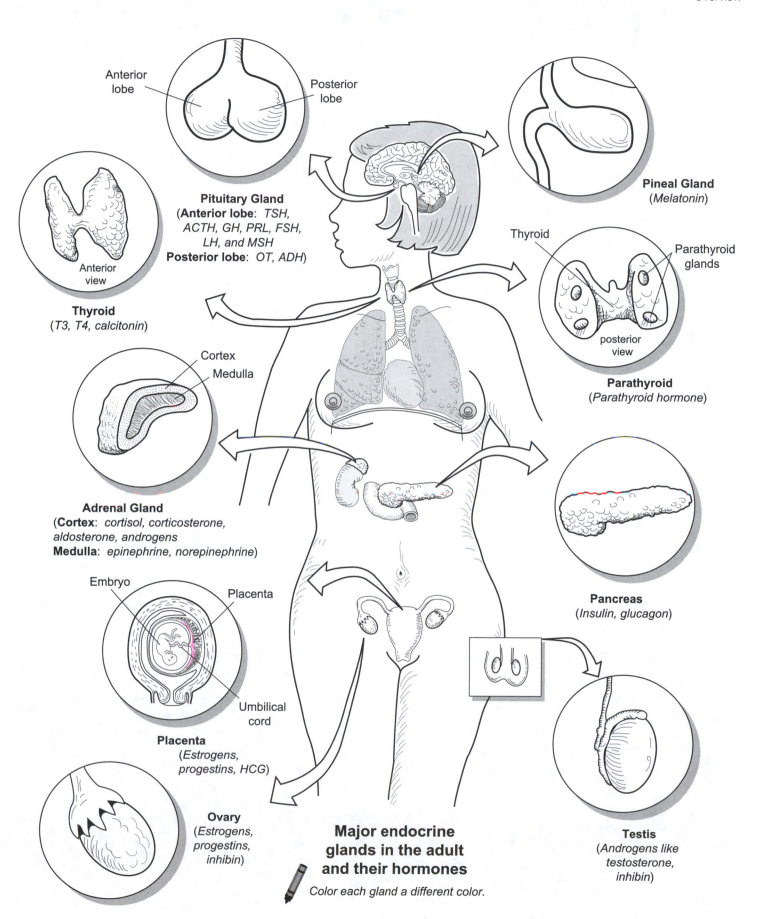

Anterior lobe

Posterior lobe

Pituitary Gland
(**Anterior lobe**: *TSH, ACTH, GH, PRL, FSH, LH, and MSH*
Posterior lobe: *OT, ADH*)

Pineal Gland
(*Melatonin*)

Anterior view

Thyroid
(*T3, T4, calcitonin*)

Thyroid

Parathyroid glands

posterior view

Parathyroid
(*Parathyroid hormone*)

Cortex

Medulla

Adrenal Gland
(**Cortex**: *cortisol, corticosterone, aldosterone, androgens*
Medulla: *epinephrine, norepinephrine*)

Pancreas
(*Insulin, glucagon*)

Embryo

Placenta

Umbilical cord

Placenta
(*Estrogens, progestins, HCG*)

Ovary
(*Estrogens, progestins, inhibin*)

Major endocrine glands in the adult and their hormones

Color each gland a different color.

Testis
(*Androgens like testosterone, inhibin*)

163

Notes

Special Senses

Description

The eyeball rests in the orbit in the skull and is surrounded by fatty tissue. Six major muscles control the movement of the eyeball:

Muscle Name	Action
Superior rectus	Elevates eye
Inferior rectus	Depresses eye
Medial rectus	Moves eye medially
Lateral rectus	Moves eye laterally
Superior oblique	Depresses eye and laterally rotates eye
Inferior oblique	Elevates eye and laterally rotates eye

The lacrimal gland produces tears that are spread across the surface of the eye during blinking. The tears accumulate in a pooling area called the **medial canthus**. Then they enter openings called the **superior lacrimal punctum** and the **inferior lacrimal punctum**, which lead to passageways called the **superior lacrimal caniculus** and the **inferior lacrimal caniculus**. From here they travel to the **lacrimal sac** and down the **nasolacrimal duct** and are drained into the nose. This explains why your nose runs when you are crying heavily.

Study Tip

The Lacrimal gland is located on the Lateral side of the eyeball. Use it as a landmark.

Key to Illustration

1. Superior lacrimal punctum (opening)
2. Superior lacrimal caniculus
3. Lacrimal sac
4. Inferior lacrimal punctum (opening)
5. Inferior lacrimal caniculus
6. Nasolacrimal duct
7. Opening of nasolacrimal duct
8. Superior oblique muscle
9. Superior rectus muscle
10. Lateral rectus muscle
11. Inferior oblique muscle
12. Inferior rectus muscle
13. Medial rectus muscle

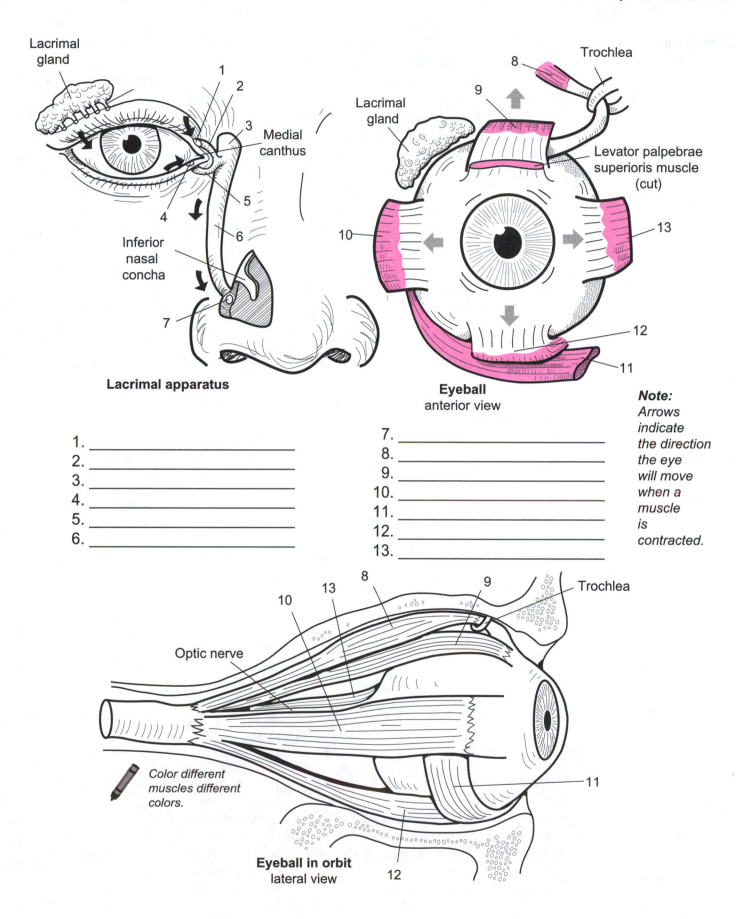

Lacrimal gland

1

2

3   Medial canthus

5

4

6

Inferior nasal concha

7

Lacrimal apparatus

Trochlea

8

9

Lacrimal gland

Levator palpebrae superioris muscle (cut)

10

13

12

11

Eyeball
anterior view

Note:
Arrows indicate the direction the eye will move when a muscle is contracted.

1. _____
2. _____
3. _____
4. _____
5. _____
6. _____

7. _____
8. _____
9. _____
10. _____
11. _____
12. _____
13. _____

8

13

10

9   Trochlea

Optic nerve

11

Color different muscles different colors.

12

Eyeball in orbit
lateral view

Description

The eye is divided into three tunics or sheaths: fibrous tunic, vascular tunic, and neural tunic. The **fibrous tunic** is a thick, tough layer of connective tissue and consists of the **sclera** (*white of the eyeball*) on its posterior portion and becomes the transparent **cornea** on its anterior portion. The cornea is cup-shaped and filled with a liquid called **aqueous humor**. The **vascular tunic** consists of the vascular, dark brown **choroid** coat on its posterior portion, which becomes the **ciliary body**, and the **iris** (*colored part of the eye*) on its anterior portion.

The lens of the eye is held in place by **suspensory ligaments** that anchor it to the ciliary body. When the smooth muscle in the ciliary body contracts, it can change the shape of the lens, allowing the eye to focus on near versus distant objects. The iris covers the front of the lens. In the center of the iris is a hole called the **pupil**, which allows light to enter the eye. Dilation and constriction of the pupil controls the amount of light that enters the eye. The **neural tunic** is the **retina** and covers only the posterior portion of the eye. This thin layer contains photoreceptors called **cones** and **rods**. Rods help us see in low-light situations, and cones help us see color and give us sharper, clearer images.

Filling the **posterior cavity** of the eye is a jelly-like substance called **vitreous humor**. It maintains the normal shape of the eyeball. On the back of the retina is a small disc called the **macula lutea**, which contains cones but no rods. At the center of this structure is a shallow depression called the **fovea centralis,** which has the highest concentration of cones and gives us our sharpest vision. In contrast, the one region of the retina that lacks photoreceptors is called the **optic disc** (*blind spot*). No images form here. It is located where the optic nerve leaves the eye.

Analogy

The **macula lutea** is like a **target** and the **fovea centralis** is like the **bullseye**.

Study Tip

To recall the order of the layers in the back of the eye, use the mnemonic: **S**cared **C**ats **R**un. This indicates the layers from outermost to innermost or **S**clera, **C**horoid, **R**etina.

Key to Illustration

1. Iris
2. Cornea
3. Edge of pupil
4. Lens
5. Anterior chamber (*filled with aqueous humor*)
6. Posterior chamber (*filled with aqueous humor*)
7. Anterior cavity
8. Suspensory ligaments
9. Ciliary body
10. Posterior cavity (*filled with vitreous humor*)
11. Central vein
12. Central artery
13. Optic nerve
14. Fovea centralis
15. Macula lutea
16. Optic disc (*blind spot*)

Color all the different layers different colors.

To recall the layers on the posterior portion of the eyeball:

| **S**cared **C**ats **R**un! | = | **S**clera **C**horoid **R**etina |

The macula lutea is like the target and the fovea centralis is the bullseye.

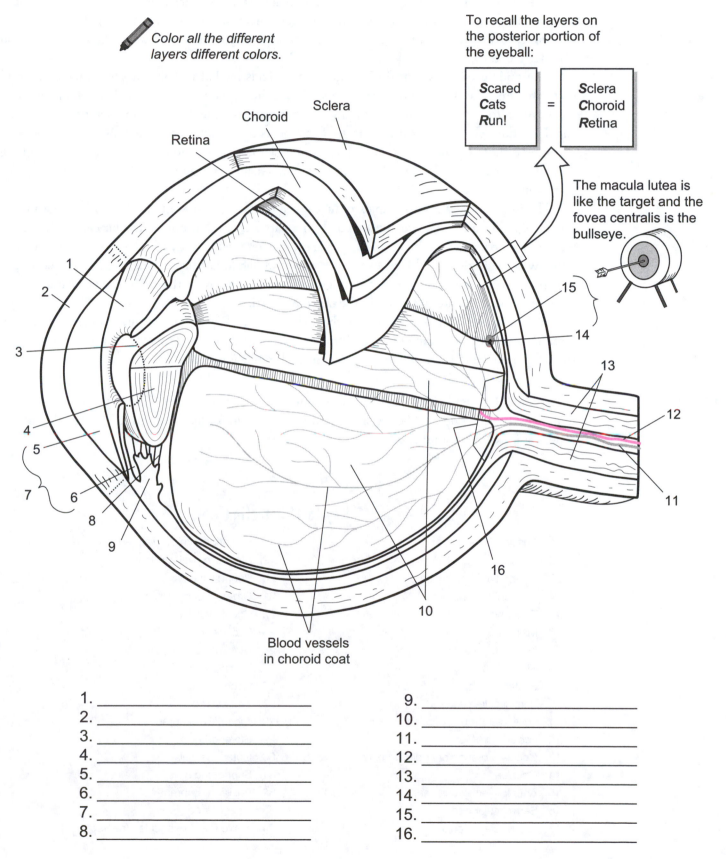

Retina

Choroid

Sclera

Blood vessels in choroid coat

1. _____
2. _____
3. _____
4. _____
5. _____
6. _____
7. _____
8. _____

9. _____
10. _____
11. _____
12. _____
13. _____
14. _____
15. _____
16. _____

Description

The ear is divided into three regional areas: external ear, middle ear, and inner ear. The **external ear** extends from the ear itself to the tympanic membrane. The **auricle** is made of elastic cartilage and directs sound waves into the **external auditory meatus.** At the end of this passageway is the delicate **tympanic membrane** or eardrum. This cone-shaped connective tissue vibrates when sound waves strike it.

The **middle ear** is an air-filled space that extends medial to the tympanic membrane and up to the inner ear. It contains three ear **ossicles: malleus, incus,** and **stapes.** These bones act as a lever system to both transmit and amplify sound waves from the tympanic membrane to the inner ear. The **auditory tube,** or **Eustachian tube,** connects the middle ear to the nasopharynx. Because this tube is short and horizontal in children it is easier for bacteria to enter the middle ear, causing middle ear infections. The function of this tube is to equalize pressure on both sides of the tympanic membrane.

The **inner ear** is a bony complex of fluid-filled chambers that contain receptors for both hearing and equilibrium. The receptors for equilibrium are located in the **semicircular canals**, and those for hearing are located in the **cochlea.** As the ear ossicles vibrate, they transfer vibration to the **oval window** of the inner ear. This sends a shock wave through fluid-filled chambers within the cochlea, which are eventually dissipated out through the **round window** to the air in the **tympanic cavity.** This vibration stimulates hair cells in the **organ of Corti.**

Each hair cell is linked to a nerve fiber. The hair cell transforms this mechanical, vibrational force into an electrical stimulus that carries nervous impulses along the cochlear nerve branch to the temporal lobe in the cerebrum, where it is interpreted as sound.

Analogy

The **tympanic membrane** looks like the **cones** on a stereo speaker. The **malleus** looks like a **hammer,** the **incus** looks like an **anvil,** and the **stapes** looks like a **stirrup.** The **cochlea** of the inner ear looks like a **snail shell.**

Key to Illustration

1. Auricle
2. Lobule
3. External auditory meatus
4. Tympanic membrane
5. Tympanic cavity
6. Malleus
7. Incus
8. Stapes
9. Eustachian (*auditory*) tube
10. Semicircular canals
11. Vestibular branch of auditory nerve
12. Facial nerve
13. Cochlear branch of auditory nerve
14. Cochlea
15. Anterior semicircular canal
16. Semicircular ducts
17. Lateral semicircular canal
18. Posterior semicircular canal
19. Stapes in oval window
20. Round window
21. Organ of Corti
22. Cochlear duct
23. Saccule
24. Utricle
25. Vestibular duct
26. Cochlear duct
27. Tympanic duct
28. Tectorial membrane
29. Stereocilia
30. Outer hair cells
31. Supporting cells
32. Basilar membrane
33. Inner hair cell
34. Nerve fibers of cochlear nerve

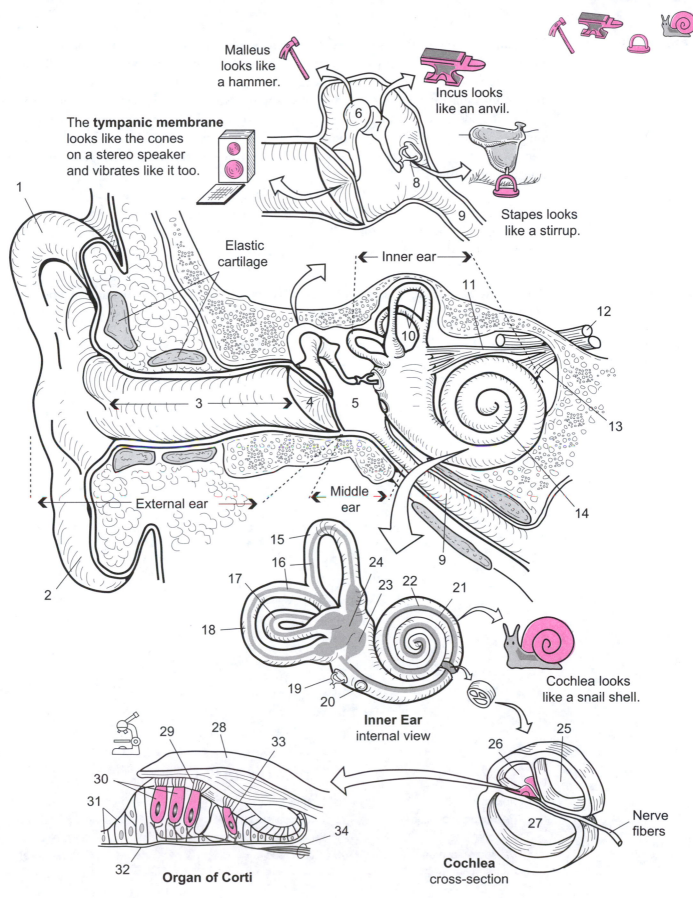

Malleus looks like a hammer.

Incus looks like an anvil.

The **tympanic membrane** looks like the cones on a stereo speaker and vibrates like it too.

Elastic cartilage

Inner ear

Stapes looks like a stirrup.

External ear

Middle ear

Inner Ear
internal view

Cochlea looks like a snail shell.

Nerve fibers

Cochlea
cross-section

Organ of Corti

171

Description

The surface of the tongue is covered with many small epithelial projections called **papillae**. There are three different types of papillae: **filiform**, **fungiform**, and **circumvallate**. The filiform papillae are located on the tip of the tongue; the fungiform papillae are located posterior to the filiform; and the **circumvallate papillae** are found in a "V" shaped strip along the posterior margin of the tongue.

Taste buds are located along the sides of the papillae. Each type of papilla contains a different number of taste buds. In total, the average adult has about 10,000 taste buds but this number decreases with age. A taste bud is composed of two different types of cells—**gustatory** (*taste*) **cells** and **supporting cells**.

The gustatory cells are modified neurons that have **microvilli** (*taste hairs*) that protrude onto the surface of a papilla. Chemicals in food bind at receptors in these microvilli, which trigger a nervous impulse in the gustatory cells. This follows a nerve pathway to the gustatory cortex in the cerebrum of the brain where the taste is interpreted.

Analogy

The **fungiform** (*fungus*, mushroom) **papilla** is dome-shaped like a **mushroom cap**, and the **filiform** (*filum*, thread) **papilla** looks like a **flame**.

Key to Illustration

1. Epiglottis	5. Lingual tonsil	9. Taste bud
2. Palatopharyngeal arch	6. Circumvallate papilla	10. Gustatory *(taste)* cells
3. Palatine tonsil	7. Fungiform papilla	11. Supporting cells
4. Palatoglossal arch	8. Filiform papilla	12. Microvilli *(taste hairs)*

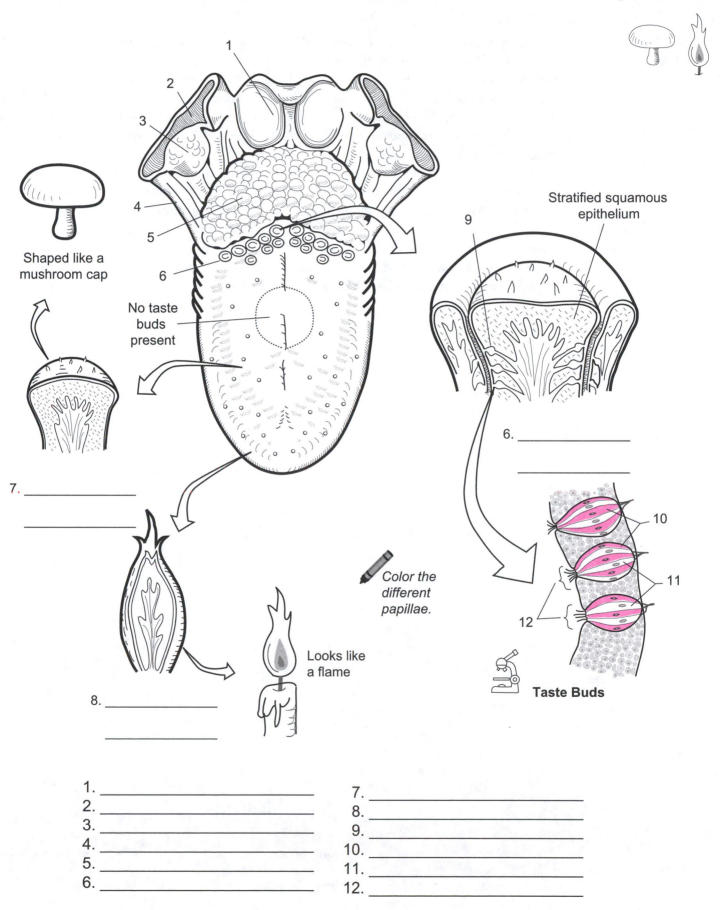

Shaped like a
mushroom cap

No taste
buds
present

Stratified squamous
epithelium

Color the
different
papillae.

Looks like
a flame

Taste Buds

7. _____

   _____

8. _____

   _____

6. _____

   _____

1. _____   7. _____
2. _____   8. _____
3. _____   9. _____
4. _____   10. _____
5. _____   11. _____
6. _____   12. _____

Description

Olfaction refers to the sense of smell. Let's follow what occurs when you smell a cup of coffee. As you inhale air through the nostrils it enters the nasal cavity and the nasal conchae produce turbulent airflow. This disperses the air and delivers aromatic molecules to the two **olfactory organs** located on the roof of the nasal cavity. These small organs are coated with a thick layer of mucus produced by the **olfactory glands**. The aromatic molecules diffuse through this mucus and bind to a receptor in the numerous **olfactory cilia**. These cilia are extensions of modified neurons called **olfactory receptor cells**.

Once the aromatic molecule binds to the receptor, it triggers a nervous impulse in the olfactory receptor cell. As this impulse travels along the cell, it passes through the **olfactory foramen** in the **cribriform plate** of the ethmoid. Then it reaches the **olfactory bulb,** which is the terminal portion of the first cranial nerve. Within the olfactory bulb, the olfactory receptor cell forms synaptic connections with other neurons. The impulse is transferred to these neurons and continues down the **olfactory tract**. Finally, the impulses are carried to the appropriate olfactory interpretation areas in the brain. These include regions in both the frontal and the temporal lobes of the cerebrum.

Key to Illustration

1. Olfactory bulb	4. Supporting cells	8. Lamina propria
2. Olfactory tract	5. Olfactory foramen	9. Olfactory epithelium
3. Olfactory receptor cells (neurons)	6. Cribriform plate of ethmoid	10. Mucous layer
	7. Olfactory gland	11. Olfactory cilia

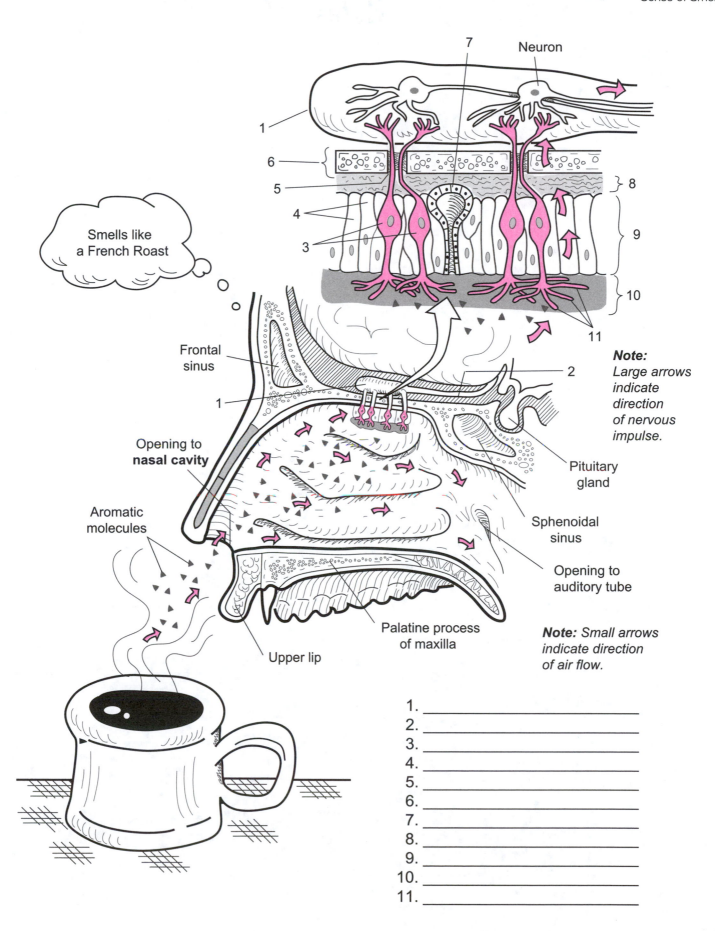

Neuron

7

1

6

5

8

4

3

9

10

11

2

Smells like
a French Roast

Frontal
sinus

1

Opening to
nasal cavity

Aromatic
molecules

Upper lip

Palatine process
of maxilla

Note:
Large arrows
indicate
direction
of nervous
impulse.

Pituitary
gland

Sphenoidal
sinus

Opening to
auditory tube

Note: Small arrows
indicate direction
of air flow.

1. _____
2. _____
3. _____
4. _____
5. _____
6. _____
7. _____
8. _____
9. _____
10. _____
11. _____

Notes

Blood

Description

Blood is a specialized type of connective tissue because it contains cells, fibers, and a liquid ground substance. It is composed of two major parts: **plasma** and **formed elements.** The **plasma** is a straw-colored fluid that contains mostly water, proteins, and other solutes. The **formed elements** consist of the following blood cells and cell fragments scattered in the plasma: **erythrocytes** (*red blood cells*), **leukocytes** (*white blood cells*), and **thrombocytes** (*platelets*).

Formed Elements of Blood

Cell	Description	Function
Erythrocytes (*red blood cells*)	Comprise 99.9% of all blood cells; biconcave discs; no nucleus in mature cell; filled with the protein hemoglobin	Transport O_2 from lungs to body cells Transport CO_2 from blood to lungs
Leukocytes (*white blood cells*)	Less than 0.1% of all blood cells; 5 different types; some have granules in cytoplasm, others do not; nucleus present in all types	Fight against pathogens such as bacteria and viruses
Thrombocytes (*platelets*)	Less than 0.1% of formed elements; cell fragments; no nucleus; contain enzymes	Involved in blood clotting

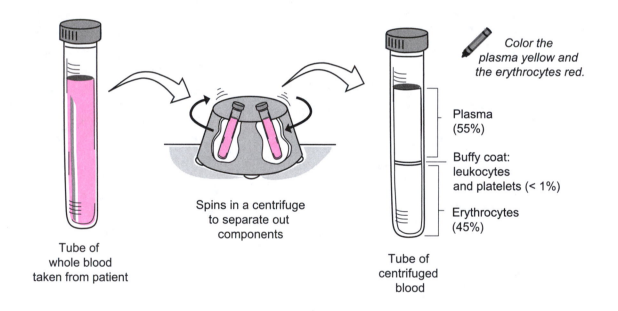

Color the plasma yellow and the erythrocytes red.

Plasma (55%)

Buffy coat: leukocytes and platelets (< 1%)

Erythrocytes (45%)

Spins in a centrifuge to separate out components

Tube of whole blood taken from patient

Tube of centrifuged blood

Key to Illustration

1. **Erythrocytes** (*red blood cells*)

Leukocytes (*white blood cells*)
2. Monocyte
3. Lymphocyte
4. Neutrophil

5. Eosinophil
6. Basophil
7. Thrombocytes (*platelets*)

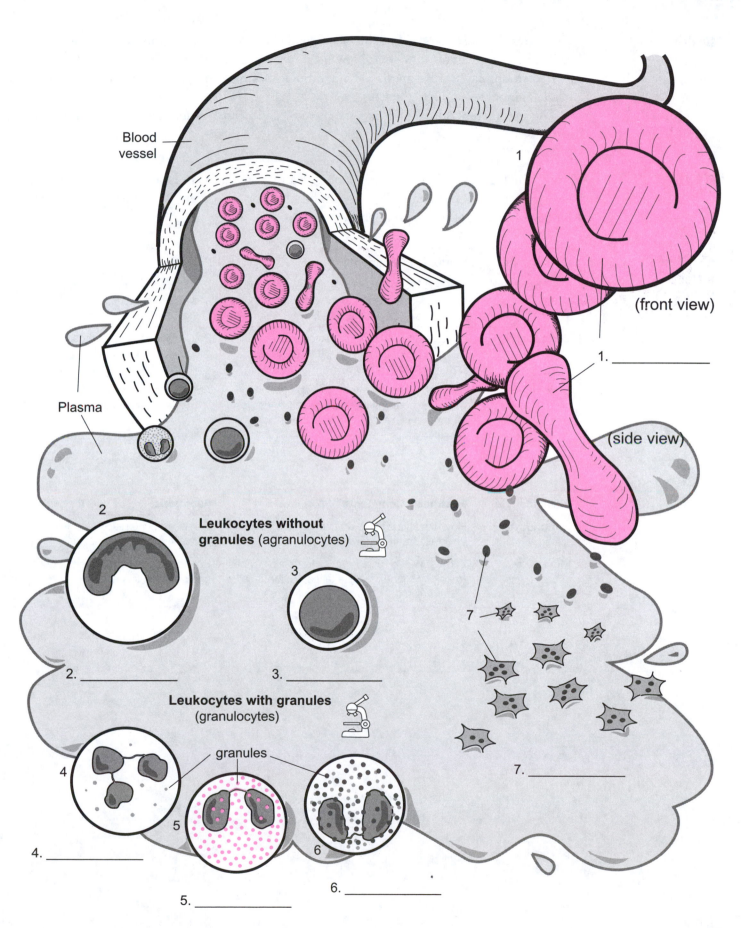

Blood vessel

Plasma

1

(front view)

1. _____

(side view)

2

Leukocytes without granules (agranulocytes)

3

2. _____

3. _____

Leukocytes with granules (granulocytes)

4

granules

5

6

7

2. _____

4. _____

5. _____

6. _____

7. _____

Description

Leukocytes (*white blood cells*, or *WBCs*) are divided into two groups: **granulocytes** and **agranulocytes**. The **granulocytes** all contain granules in the cytoplasm of the cell and include **neutrophils**, **eosinophils**, and **basophils**. The **agranulocytes** have no granules in the cytoplasm and include **monocytes** and **lymphocytes**.

The general function of all leukocytes is to defend against various pathogens such as bacteria and viruses.

Study Tip

To rank the leukocytes from most common to least common, use the following mnemonic: **N**ever **L**et **M**onkeys **E**at **B**ananas. This gives you the correct order: **N**eutrophils, **L**ymphocytes, **M**onocytes, **E**osinophils, and **B**asophils.

Features and Functions of WBCs

Granulocyte	Features / Comments	Functions
Neutrophil	Nucleus has 3–5 lobes; contains least amount of granules; named after the fact that granules are *neutral*—do not stain well.	Phagocytic cell; engulfs bacteria and debris in tissues.
Eosinophil	Nucleus usually has 2 lobes; large granules that stain brightly; named after dye used to stain granules—*eosin* dye.	Phagocytic cell; fights parasitic infections; engulfs anything labeled with antibodies; reduces inflammation.
Basophil	Nucleus usually masked by deep purple/blue granules; contains the most granules; named after the *basic* stain used to stain granules—hematoxylin.	Assists in damaged tissue repair by releasing histamine from granules.

Agranulocyte	Features / Comments	Functions
Lymphocyte	Round nucleus takes up nearly entire cell volume; slightly larger than RBC.	Part of immune response; defend against pathogens or toxins.
Monocyte	Largest WBC; nucleus varies from horseshoe to kidney shape; nucleus takes up about half of cell volume.	Phagocytic cell; engulfs pathogens and debris in tissues.

Ranking the Leukocytes from Most Common to Least Common

My Drawing of the WBCs

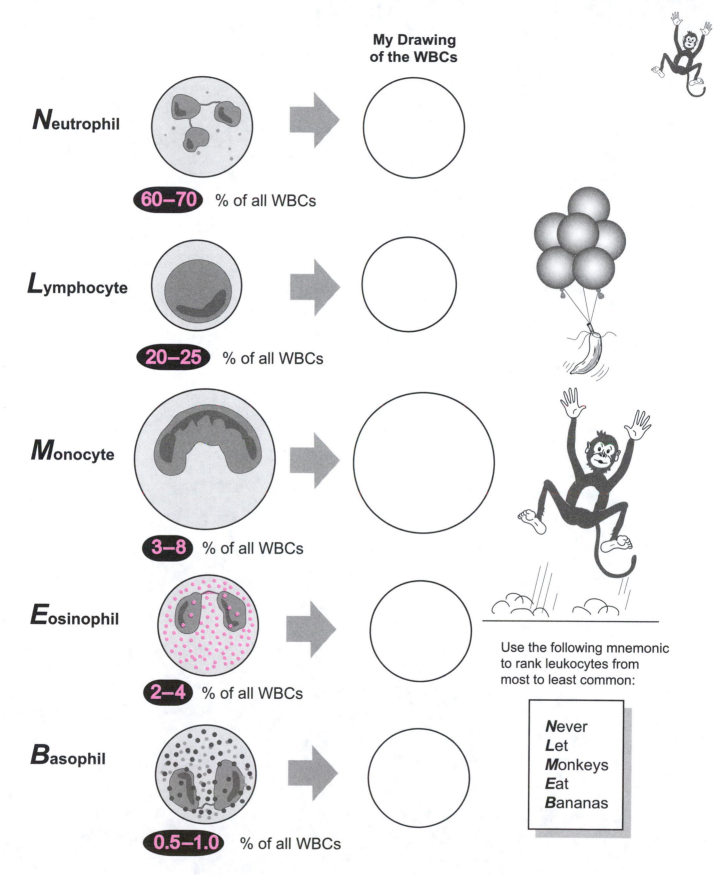

Neutrophil

60–70 % of all WBCs

Lymphocyte

20–25 % of all WBCs

Monocyte

3–8 % of all WBCs

Eosinophil

2–4 % of all WBCs

Basophil

0.5–1.0 % of all WBCs

Use the following mnemonic to rank leukocytes from most to least common:

Never
Let
Monkeys
Eat
Bananas

Notes

Cardiovascular System

Description

Coronary circulation refers to the blood supply to the heart. The **coronary arteries** supply oxygenated blood to the heart and the **cardiac veins** carry deoxygenated blood back to the heart. The following flowchart summarizes coronary circulation through the blood vessels.

> **Base of aorta** ⟶ left and right coronary arteries ⟶ branches of coronary arteries (*circumflex a., anterior interventricular a., marginal a., posterior interventricular a.*) ⟶ capillaries ⟶ cardiac veins ⟶ coronary sinus ⟶ **right atrium**

When **coronary arteries** become blocked, the blood supply to the heart is reduced. This deprives cardiac muscle cells of oxygen. If this blockage persists over many years, it may lead to a **myocardial infarction** (*heart attack*).

Analogy

A **sulcus** is like a **shallow groove** or **gulley**.

Study Tip

To distinguish the anterior from the posterior view of the heart, use the **coronary sinus** as a landmark for the posterior view. It is often difficult to see on a dissected specimen because it is normally covered with a horizontal band of fatty tissue. Good landmarks for the anterior view include the **pulmonary trunk**, **anterior interventricular artery**, **circumflex artery**, and **ascending aorta**.

Key to Illustration

Blood vessels (B)

B1. Superior vena cava
B2. Inferior vena cava
B3. Ascending aorta
B4. Aortic arch
B5. Descending aorta
B6. Brachiocephalic trunk
B7. L. Common carotid a.
B8. L. Subclavian a.
B9. Pulmonary trunk
B10. Pulmonary arteries
B11. Pulmonary veins
B12. R. Coronary a. *(in r. anterior atrioventricular groove)*
B13. L. Coronary a.

B14. Circumflex a.
B15. Anterior Interventricular a. *(in anterior interventricular sulcus)*
B16. Marginal a.
B17. Anterior cardiac v.
B18. Great cardiac v.
B19. Small cardiac v.
B20. Coronary sinus
B21. Posterior v. of l. ventricle
B22. Middle cardiac v.
B23. Posterior interventricular a. *(in posterior interventricular sulcus)*

Structures (S)

S1. Ligamentum arteriosum
S2. Apex *(tip)* of heart
S3. Auricle

Chambers (C)

C1. R. atrium
C2. L. atrium
C3. R. ventricle
C4. L. ventricle

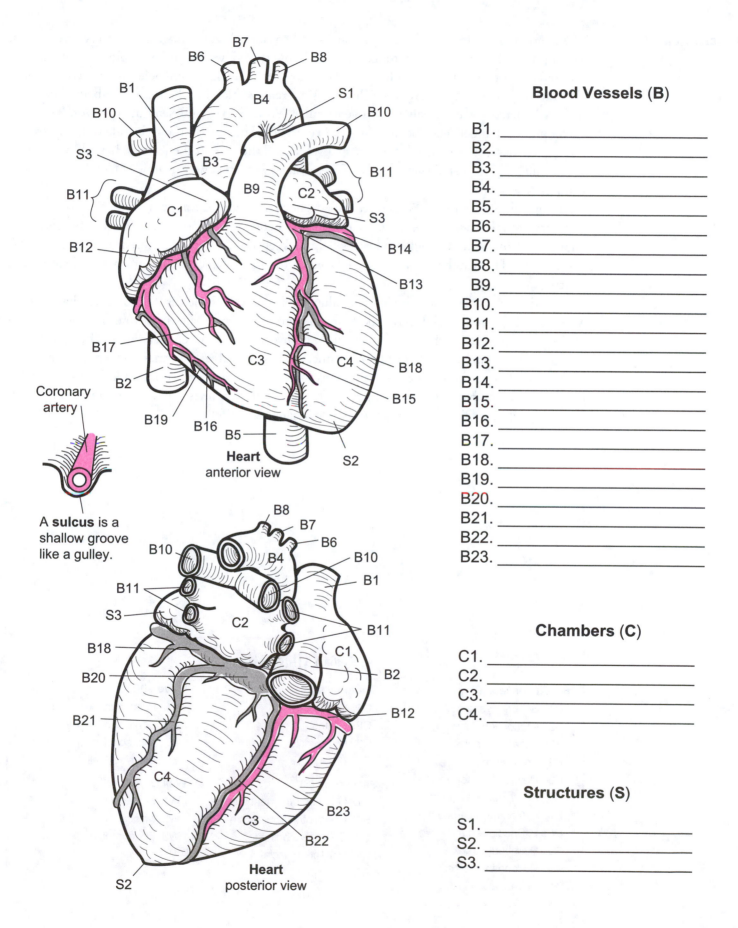

Coronary artery

A **sulcus** is a shallow groove like a gulley.

Heart
anterior view

Heart
posterior view

Blood Vessels (B)

B1. _____
B2. _____
B3. _____
B4. _____
B5. _____
B6. _____
B7. _____
B8. _____
B9. _____
B10. _____
B11. _____
B12. _____
B13. _____
B14. _____
B15. _____
B16. _____
B17. _____
B18. _____
B19. _____
B20. _____
B21. _____
B22. _____
B23. _____

Chambers (C)

C1. _____
C2. _____
C3. _____
C4. _____

Structures (S)

S1. _____
S2. _____
S3. _____

Description

The heart is divided into left and right halves and has four chambers, two atria and two ventricles. The **atria** are the first chambers to receive blood from the body. They fill with blood, contract, and transfer blood to the pumping chambers or **ventricles**. The **right ventricle** pumps deoxygenated blood to the lungs and the **left ventricle** pumps oxygenated blood to the rest of the body. The heart has two different types of valves: **atrioventricular (A-V) valves** and **semilunar valves**. The A-V valves are located between the atria and the ventricles. The one on the right side of the heart has three valve flaps, so it is called the **tricuspid valve**, and the one on the left side has two valve flaps so it is called the **bicuspid** (*mitral*) **valve**. These valves permit a one-way flow of blood from atria to ventricles.

Long, fibrous, cord-like structures called **chordae tendineae** anchor the valve flaps to the **papillary muscles**, which are long, cone-shaped muscular extensions of the inner ventricles. The chordae tendineae and papillary muscles help to keep the A-V valves closed during ventricular contraction. The semilunar valves are located at the base of each major artery that leaves each ventricle.

On the right side is the pulmonary semilunar valve and on the left is the aortic semilunar valve. These valves prevent backflow of blood into the ventricles. From outermost to innermost, the wall of the heart is made of three layers: epicardium, myocardium, and endocardium. The **epicardium** (*visceral pericardium*) is made of fibrous connective tissue and is the innermost layer of the pericardial sac that surrounds the heart. The **myocardium** is composed of multiple layers of cardiac muscle and many blood vessels and nerves. The **endocardium** lines the inside of all the chambers along with all the valves and is made of simple squamous epithelium.

Key to Illustration

Blood Vessels (B)
B1. Superior vena cava
B2. Inferior vena cava
B3. Ascending aorta
B4. Aortic arch
B5. Descending aorta
B6. Brachiocephalic trunk
B7. L. Common carotid a.
B8. L. Subclavian a.
B9. Pulmonary trunk
B10. Pulmonary arteries
B11. Pulmonary veins

Chambers (C)
C1. R. atrium
C2. L. atrium
C3. R. ventricle
C4. L. ventricle

Structures (S)
S1. Ligamentum arteriosum
S2. Interventricular septum
S3. Chordae tendineae
S4. Papillary muscle
S5. Apex *(tip)*

Valves (V)
V1. Pulmonary semilunar
V2. Aortic semilunar
V3. Tricuspid leaflet
V4. Bicuspid leaflet

Wall Layers (W)
W1. Epicardium
W2. Myocardium
W3. Endocardium

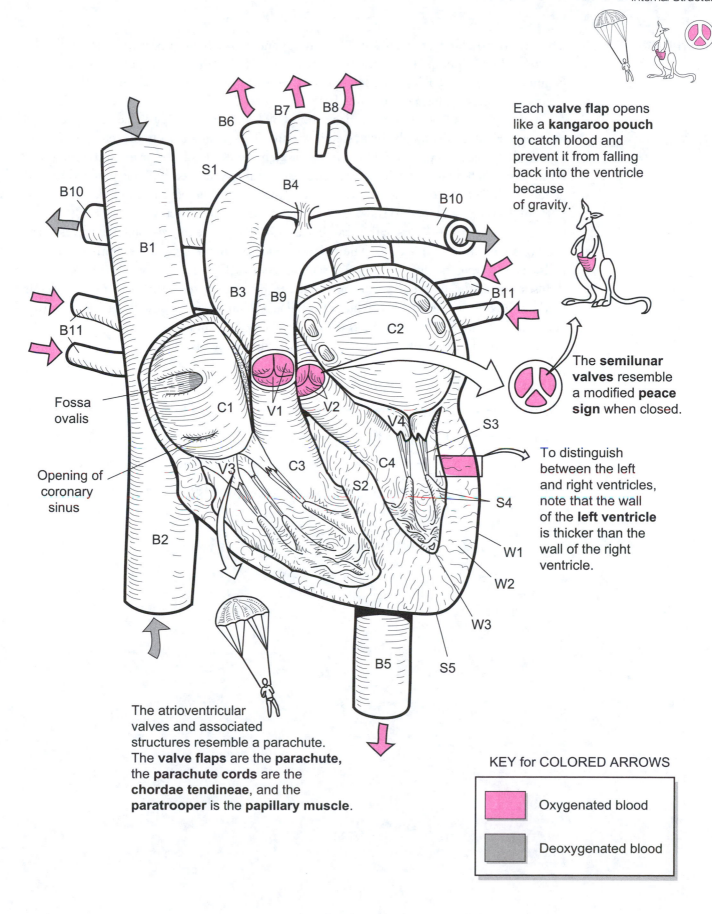

Each **valve flap** opens like a **kangaroo pouch** to catch blood and prevent it from falling back into the ventricle because of gravity.

The **semilunar valves** resemble a modified **peace sign** when closed.

To distinguish between the left and right ventricles, note that the wall of the **left ventricle** is thicker than the wall of the right ventricle.

B6 B7 B8

S1

B4

B10

B10

B1

B3 B9

C2

B11

B11

Fossa ovalis

C1

V1 V2

V4

S3

Opening of coronary sinus

V3

C3

C4

S2

S4

W1

B2

W2

W3

B5

S5

The atrioventricular valves and associated structures resemble a parachute. The **valve flaps** are the **parachute,** the **parachute cords** are the **chordae tendineae**, and the **paratrooper** is the **papillary muscle**.

KEY for COLORED ARROWS

▮ (pink)	Oxygenated blood
▮ (gray)	Deoxygenated blood

Description

The **cardiovascular system** consists of the **heart** and all the **blood vessels**. Functionally, the heart is like a double pump. It consists of two receiving chambers called **atria** and two pumping chambers called **ventricles**. The left side of the heart always pumps oxygenated blood while the right side always pumps deoxygenated blood.

The illustration on the facing page shows blood flow through the heart, through the **pulmonary circuit** and through the **systemic circuit**. The pulmonary circuit refers to all the blood vessels that take deoxygenated blood from the right ventricle of the heart to the lungs and then return oxygenated blood to the left atrium. After this oxygenated blood is pumped from the left atrium to the left ventricle, it is pumped out to the rest of the body. The blood vessels that transport this oxygenated blood to the body are part of the systemic circuit.

All gas exchange occurs within **capillaries**. Capillaries are microscopic blood vessels that are only one cell layer in thickness. Their wall is made of **simple squamous epithelium**. These flat cells easily permit the diffusion of gases such as **oxygen (O_2)** and **carbon dioxide (CO_2)**. Oxygen diffuses out of the blood and into body cells to be used in the process of cellular respiration. Carbon dioxide is a normal byproduct of cellular respiration and gradually builds up within body cells. Carbon dioxide diffuses from the body cells into the capillary.

Beginning in the right atrium, this is a flowchart for the blood flow:

Right atrium (1) → tricuspid valve (2) → right ventricle (3) → pulmonary semilunar valve (4) → pulmonary trunk (5) → pulmonary arteries (6) → lungs → pulmonary veins (7) → left atrium (8) → bicuspid valve (9) → left ventricle (10) → aortic semilunar valve (11) → ascending aorta (12) → aortic arch (13) → descending aorta (14) → inferior vena cava (15) and superior vena cava (16) → right atrium (1)

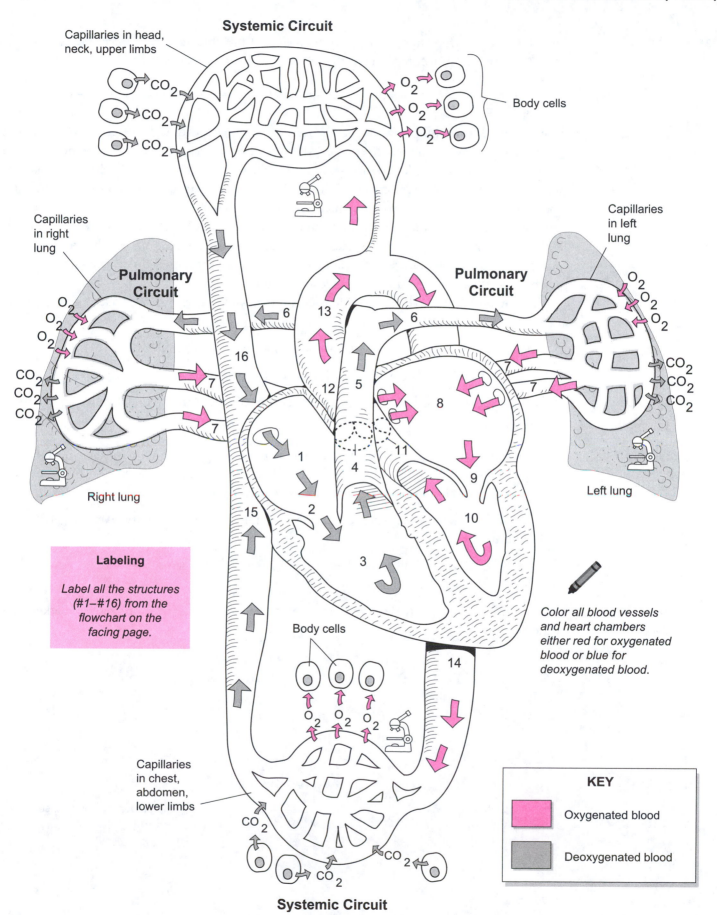

Systemic Circuit

Capillaries in head, neck, upper limbs

CO$_2$
CO$_2$
CO$_2$

O$_2$
O$_2$
O$_2$

Body cells

Capillaries in right lung

Capillaries in left lung

Pulmonary Circuit

Pulmonary Circuit

O$_2$
O$_2$
O$_2$

O$_2$ O$_2$ O$_2$

CO$_2$
CO$_2$
CO$_2$

CO$_2$
CO$_2$
CO$_2$

Right lung

Left lung

13

6

6

16

7

7

12

5

8

7

7

1

11

4

9

2

10

15

3

Labeling

Label all the structures (#1–#16) from the flowchart on the facing page.

Body cells

O$_2$
O$_2$
O$_2$

14

Color all blood vessels and heart chambers either red for oxygenated blood or blue for deoxygenated blood.

Capillaries in chest, abdomen, lower limbs

CO$_2$

CO$_2$

CO$_2$

Systemic Circuit

KEY	
	Oxygenated blood
	Deoxygenated blood

Description

The body has five fundamental types of blood vessels: *arteries*, *arterioles*, *capillaries*, *venules*, and *veins*. All of them connect together in the following predictable pattern:

heart ⟶ artery ⟶ arteriole ⟶ capillary ⟶ venule ⟶ vein ⟶ **heart**

Arteries always carry blood away from the heart. They are thicker-walled than veins because the blood within them is at a higher pressure. All **veins** always carry blood back to the heart. Because the pressure within them is lower, they are thinner-walled. Larger veins contain valves at regular intervals to assist the blood to return to the heart. Arteries and veins connect together at the microscopic level by capillary networks. **Capillaries** are the smallest blood vessels in the body and are important functionally because gas and fluid exchange occurs here. Their entire wall is often a single cell layer in thickness. **Arterioles** and **venules** are microscopic vessels that feed and drain capillaries, respectively. Nearest the capillaries, they are structurally similar to a capillary except they have small amounts of smooth muscle around them.

Arteries and veins have three major layers in their walls: *tunica externa*, *tunica media*, and *tunica interna*:

- **Tunica externa**—connective tissue layer made mostly of collagen fibers

- **Tunica media**—layers of smooth muscle and some elastic fibers

- **Tunica interna**—endothelial layer (simple squamous epithelium) with underlying loose connective tissue

Study Tips

- Don't confuse yourself by trying to distinguish between arteries and veins as to whether they carry oxygenated blood or deoxygenated blood. *This does not work* because some arteries/veins carry oxygenated blood and others carry deoxygenated blood.

- To recall the general function of arteries use the phrase: *Arteries Away!* **Arteries** always carry blood *away* from the heart.

- To recall one structural difference between arteries and veins: *Veins have Valves* (arteries do not have valves).

- Under the microscope, to distinguish an artery from a vein, **arteries** always have a **thicker tunica media**.

Key to Illustration

1. Tunica interna *(tunica intima)*
2. Tunica media
3. Tunica externa *(tunica adventitia)*
4. Lumen of blood vessel

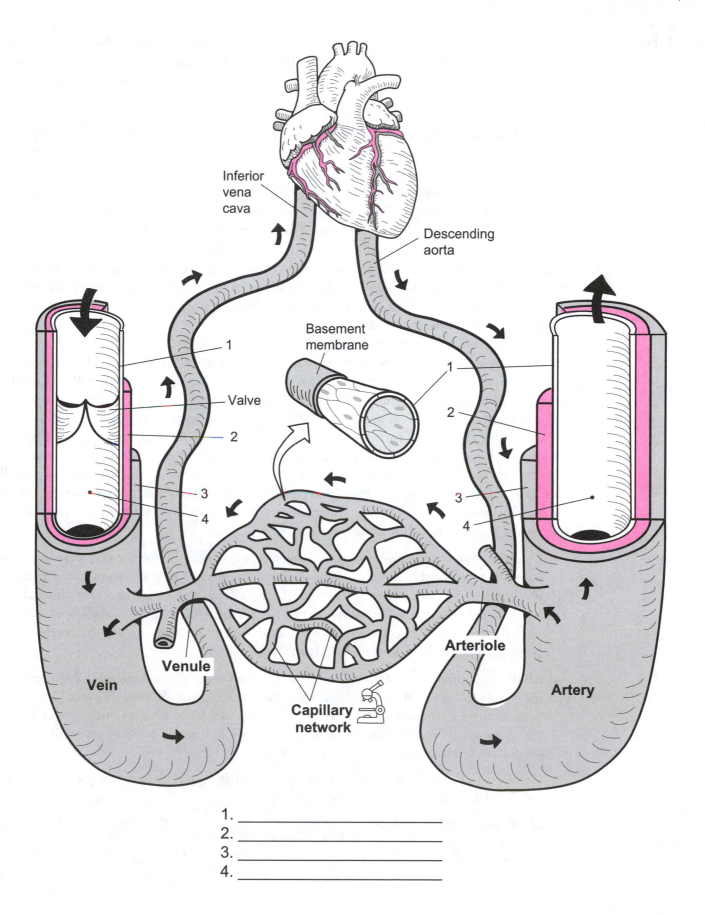

Inferior vena cava

Descending aorta

Basement membrane

Valve

Capillary network

Venule

Arteriole

Vein

Artery

1. _____
2. _____
3. _____
4. _____

Description
The illustration gives an overview of the general pattern of circulation. Blood always follows a predictable circuit through blood vessels. There are five fundamental types of blood vessels in the body: *arteries, arterioles, capillaries, venules,* and *veins.* All of them connect together in the following pattern:

Heart ⟶ artery ⟶ arteriole ⟶ capillary ⟶ venule ⟶ vein ⟶ heart

The schematic illustration on the facing page shows three of these five: *arteries, veins,* and *capillaries.* **Arteries** always carry blood away from the heart. They are thicker-walled than veins because the blood within them is at a higher pressure. As distance from the heart increases, pressure decreases. All **veins** always carry blood back to the heart. Since the pressure within them is lower, they are thinner walled. Arteries and veins connect together at the microscopic level by capillary networks. **Capillaries** are the smallest blood vessels in the body and are very important functionally since gas exchange and fluid exchange occurs here. Oxygen exits the blood to be used by body cells, and carbon dioxide enters the blood from cells. The liquid plasma is filtered out of the blood to become tissue fluid.

Let's follow the general pattern of circulation. Veins carrying low pressure, deoxygenated blood drain into the **vena cava**, which drains into the heart's **right atrium (RA)**. This receiving chamber fills with blood, contracts, and forces blood into the **right ventricle (RV)**. All this deoxygenated blood is then pumped out of the right ventricle to go to the lungs to get oxygenated. In the lungs, oxygen diffuses into the blood through the **pulmonary capillaries**. The oxygenated blood is then transported through veins to the **left atrium (LA)**. The LA fills with blood, contracts, and forces blood into the **left ventricle (LV)**. This oxygenated blood is then pumped out to the body via the **aorta**. The heart feeds its own cardiac muscle first through **coronary capillaries** so it can continue pumping blood every minute of every day. Arteries carry oxygenated blood above the heart to the capillaries in the brain, trunk, and upper limbs. Other arteries carry blood below the heart to the following major areas:

- **Digestive organs and spleen**—After gas exchange occurs at the **splenic** and **mesenteric capillaries**, deoxygenated blood is carried by veins to the **hepatic portal system** in the liver. Note that capillaries in this system are not for the typical purpose of gas exchange. Instead, these highly permeable capillaries are specialized for delivering nutrients absorbed by the digestive tract to liver cells. The liver cells serve as special processing centers that perform many functions. For example, they detoxify harmful substances.

- **Kidney**—Another unique group of permeable capillaries is the **glomerular capillaries**. Like the capillaries in the hepatic portal system, these are also not for the purpose of gas exchange. Instead, they are specialized to filter the blood plasma, place it in a separate tubular system, and process this liquid into urine. These capillaries lead into the **peritubular capillaries** where gas exchange does occur.

- **Gonads**—In the male, gas exchange occurs at capillaries in the testes whereas, in the female, gas exchange occurs at capillaries in the ovaries.

- **Liver, lower limbs**

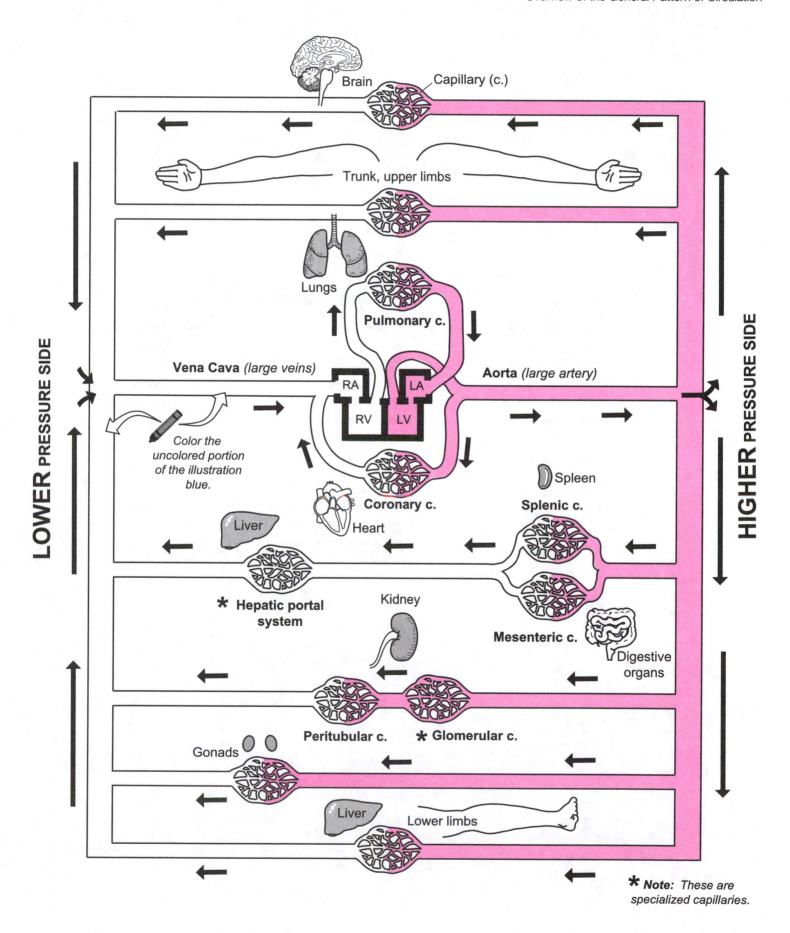

Brain

Capillary (c.)

Trunk, upper limbs

Lungs

Pulmonary c.

Vena Cava (*large veins*)

Aorta (*large artery*)

RA LA

RV LV

Color the uncolored portion of the illustration blue.

Liver

Coronary c.

Heart

Spleen

Splenic c.

LOWER PRESSURE SIDE

HIGHER PRESSURE SIDE

✱ **Hepatic portal system**

Kidney

Mesenteric c.

Digestive organs

Gonads

Peritubular c.   ✱ **Glomerular c.**

Liver

Lower limbs

✱ **Note:** These are specialized capillaries.

Key to Illustration

Head/Neck (H)

H1. Internal carotid a.

H2. External carotid a.

H3. Common carotid a.

H4. Subclavian a.

Shoulder (S)

S1. Axillary a.

Thorax (T)

T1. Aortic arch

T2. Pulmonary trunk

T3. Pulmonary a.

Arm (AR)

AR1. Brachial a.

Forearm (FO)

FO1. Ulnar a.

FO2. Radial a.

Abdomen (A)

A1. Abdominal aorta

A2. Celiac trunk

A3. Superior mesenteric a.

A4. Renal a.

A5. Gonadal a.
(testicular a. in males, ovarian a. in females)

A6. Inferior mesenteric a.

A7. Common iliac a.

A8. Internal iliac a.

A9. External iliac a.

Thigh (TH)

TH1. Femoral a.

Leg (L)

L1. Popliteal a.

L2. Anterior tibial a.

L3. Fibular a.

L4. Posterior tibial a.

Major Arteries

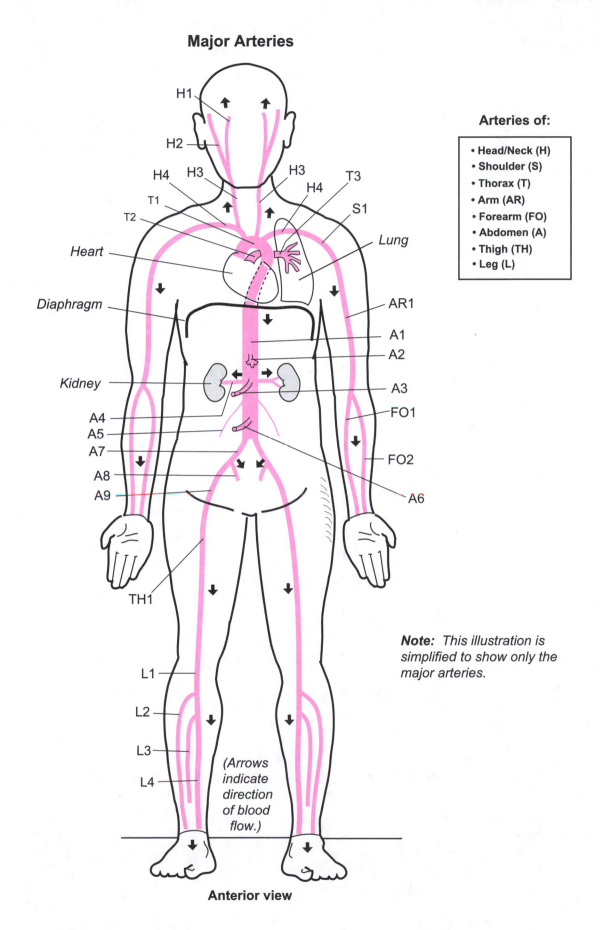

Arteries of:

- Head/Neck (H)
- Shoulder (S)
- Thorax (T)
- Arm (AR)
- Forearm (FO)
- Abdomen (A)
- Thigh (TH)
- Leg (L)

H1
H2
H4
H3
H3
H4
T3
T1
T2
S1
Heart
Lung
Diaphragm
AR1
A1
A2
Kidney
A3
A4
FO1
A5
A7
FO2
A8
A9
A6
TH1
L1
L2
L3
L4

(Arrows indicate direction of blood flow.)

Note: *This illustration is simplified to show only the major arteries.*

Anterior view

Key to Illustration

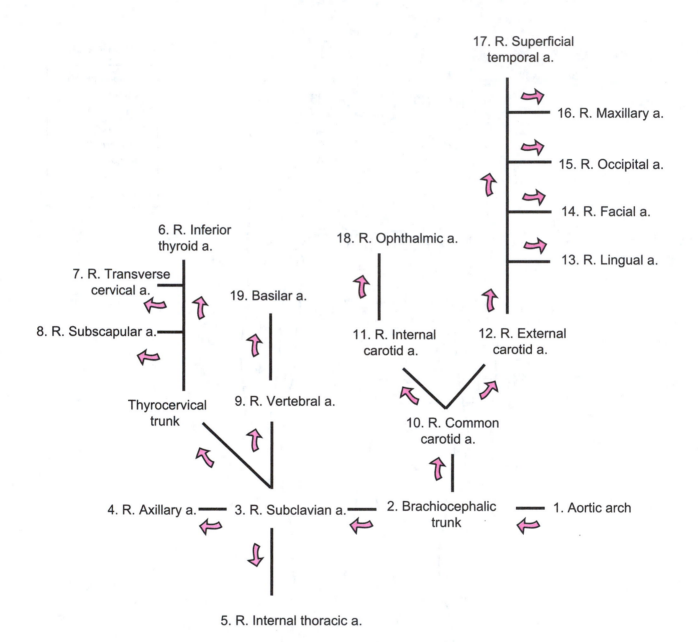

17. R. Superficial temporal a.

16. R. Maxillary a.

15. R. Occipital a.

14. R. Facial a.

13. R. Lingual a.

6. R. Inferior thyroid a.

18. R. Ophthalmic a.

7. R. Transverse cervical a.

19. Basilar a.

8. R. Subscapular a.

11. R. Internal carotid a.

12. R. External carotid a.

Thyrocervical trunk

9. R. Vertebral a.

10. R. Common carotid a.

4. R. Axillary a.

3. R. Subclavian a.

2. Brachiocephalic trunk

1. Aortic arch

5. R. Internal thoracic a.

Note: *Colored arrows indicate direction of blood flow.*

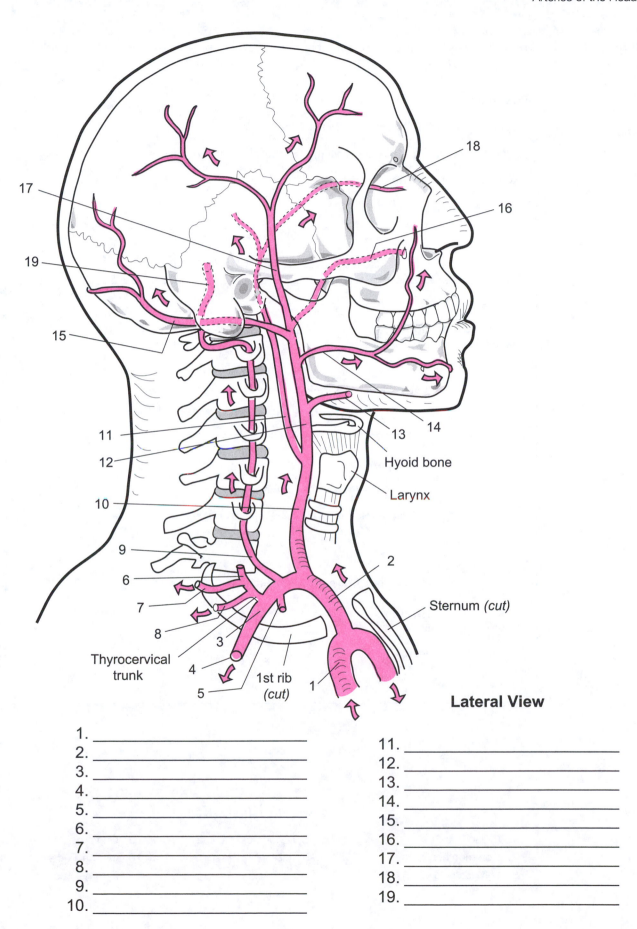

18

17

16

19

15

11

12

10

13

14

Hyoid bone

Larynx

9

6

2

7

Sternum *(cut)*

8

3

Thyrocervical
trunk

4

5

1st rib
(cut)

1

Lateral View

1. _____
2. _____
3. _____
4. _____
5. _____
6. _____
7. _____
8. _____
9. _____
10. _____

11. _____
12. _____
13. _____
14. _____
15. _____
16. _____
17. _____
18. _____
19. _____

Description   The blood vessel branches of the abdominal aorta supply oxygenated blood to the organs and structures within the abdominal cavity and the lower limbs.

Study Tips
- Use diaphragm as a landmark
- Celiac trunk is first branch below the diaphragm
- Superior mesenteric is larger in diameter than the inferior mesenteric
- Renal and gonadal are paired
- Gonadal arteries are slender

Schematic of major branches

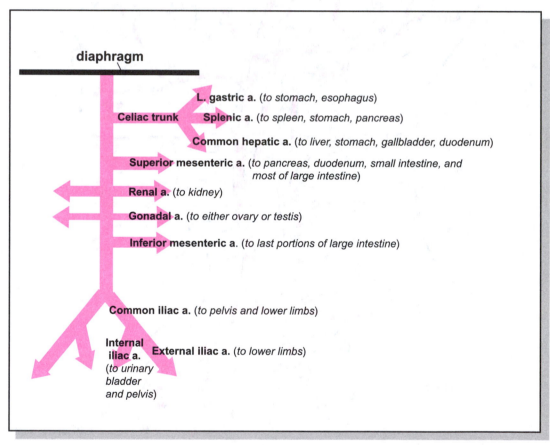

diaphragm

L. gastric a. (*to stomach, esophagus*)

Celiac trunk   **Splenic a.** (*to spleen, stomach, pancreas*)

Common hepatic a. (*to liver, stomach, gallbladder, duodenum*)

Superior mesenteric a. (*to pancreas, duodenum, small intestine, and most of large intestine*)

Renal a. (*to kidney*)

Gonadal a. (*to either ovary or testis*)

Inferior mesenteric a. (*to last portions of large intestine*)

Common iliac a. (*to pelvis and lower limbs*)

Internal iliac a. (*to urinary bladder and pelvis*)   **External iliac a.** (*to lower limbs*)

Key to Illustration

1. Celiac truck
2. Common hepatic a.
3. Left gastric a.
4. Splenic a.
5. Superior mesenteric a.
6. Renal a.
7. Gonadal a.
8. Inferior mesenteric a.
9. Common iliac a.
10. Internal iliac a.
11. External iliac a.

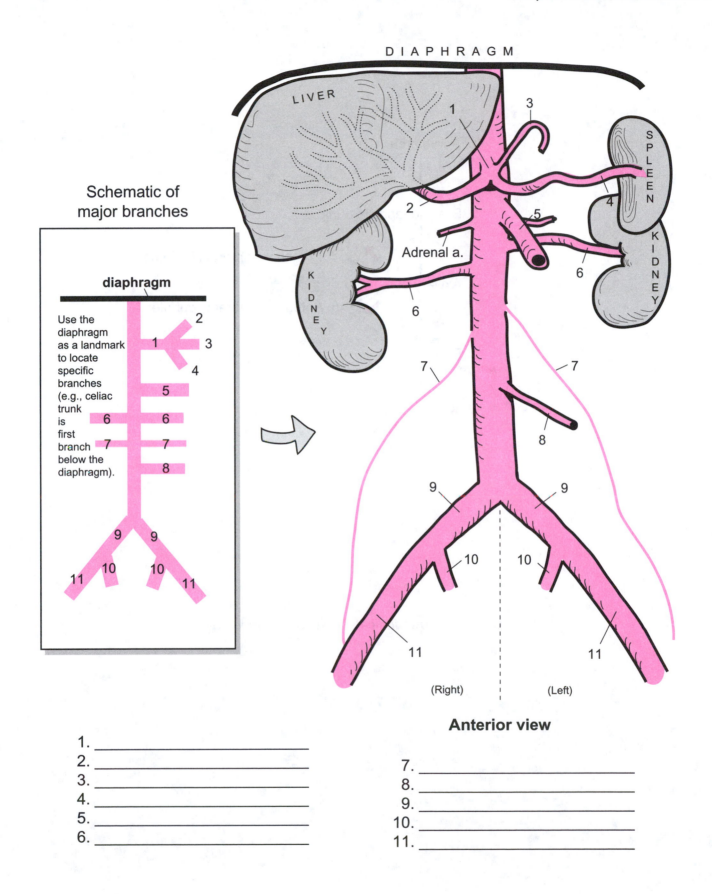

DIAPHRAGM

LIVER

SPLEEN

KIDNEY

KIDNEY

Adrenal a.

Schematic of major branches

diaphragm

Use the diaphragm as a landmark to locate specific branches (e.g., celiac trunk is first branch below the diaphragm).

(Right)   (Left)

Anterior view

1. _____
2. _____
3. _____
4. _____
5. _____
6. _____

7. _____
8. _____
9. _____
10. _____
11. _____

Key to Illustration

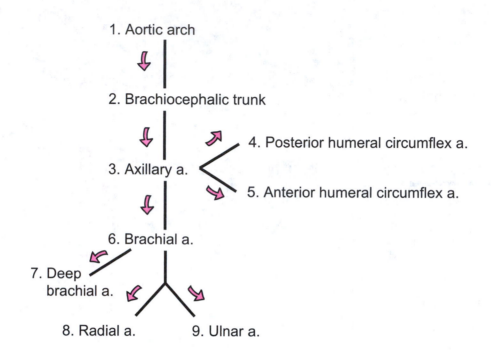

1. Aortic arch

2. Brachiocephalic trunk

3. Axillary a.

4. Posterior humeral circumflex a.

5. Anterior humeral circumflex a.

6. Brachial a.

7. Deep brachial a.

8. Radial a.

9. Ulnar a.

Note: Colored arrows indicate direction of blood flow.

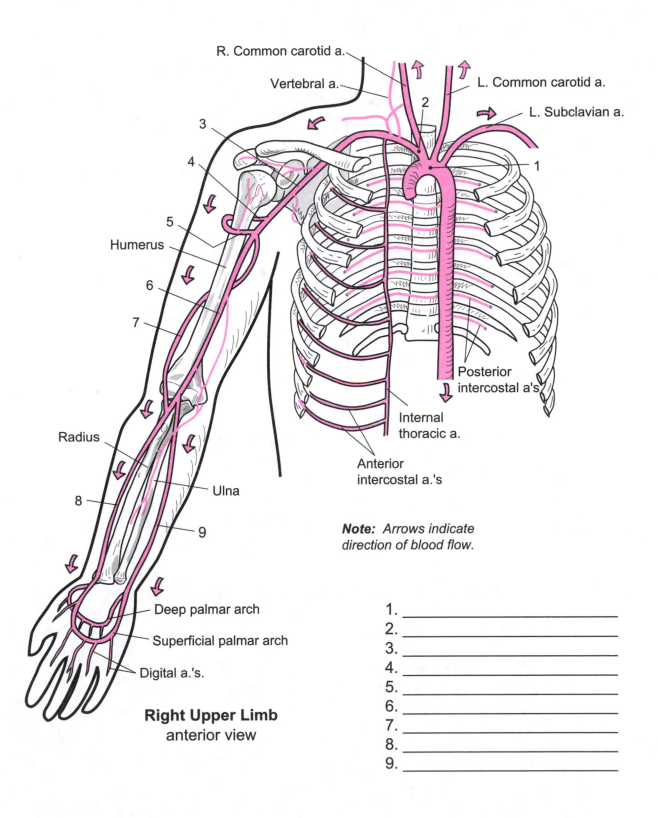

R. Common carotid a.

Vertebral a.

L. Common carotid a.

L. Subclavian a.

3

4

5

Humerus

6

7

Radius

8

Ulna

9

Posterior intercostal a's

Internal thoracic a.

Anterior intercostal a.'s

Note: *Arrows indicate direction of blood flow.*

Deep palmar arch

Superficial palmar arch

Digital a.'s.

Right Upper Limb
anterior view

1. _____
2. _____
3. _____
4. _____
5. _____
6. _____
7. _____
8. _____
9. _____

Key to Illustration

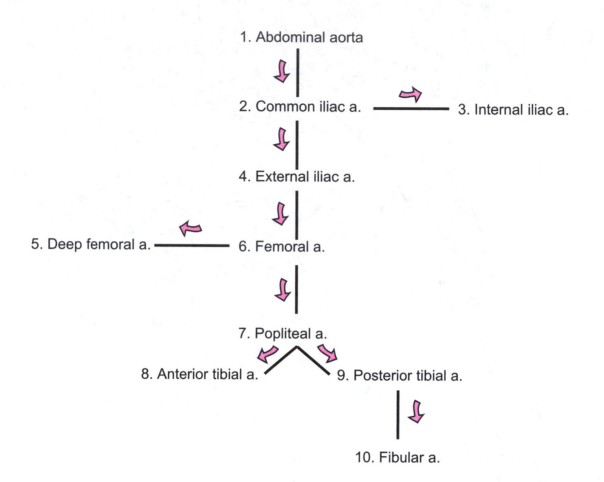

1. Abdominal aorta
2. Common iliac a.
3. Internal iliac a.
4. External iliac a.
5. Deep femoral a.
6. Femoral a.
7. Popliteal a.
8. Anterior tibial a.
9. Posterior tibial a.
10. Fibular a.

Note: Colored arrows indicate direction of blood flow.

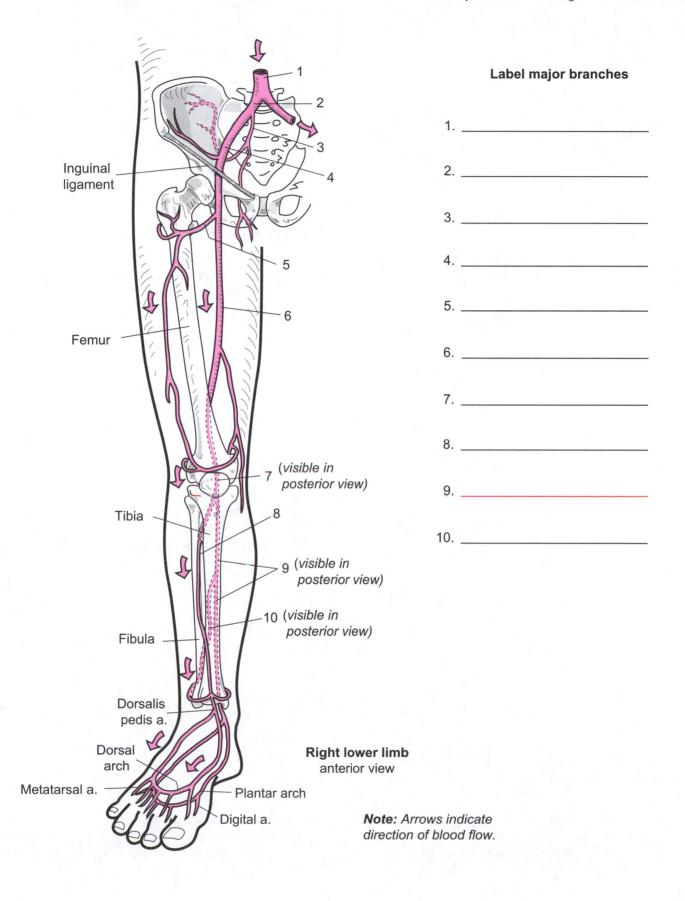

Inguinal
ligament

Femur

Tibia

Fibula

Dorsalis
pedis a.

Dorsal
arch

Metatarsal a.

Plantar arch

Digital a.

7 (*visible in
posterior view*)

8

9 (*visible in
posterior view*)

10 (*visible in
posterior view*)

Right lower limb
anterior view

Note: *Arrows indicate
direction of blood flow.*

Label major branches

1. _____

2. _____

3. _____

4. _____

5. _____

6. _____

7. _____

8. _____

9. _____

10. _____

Key to Illustration

Head/Neck (H)

H1. External jugular v.

H2. Internal jugular v.

Shoulder (S)

S1. Axillary v.

Thorax (T)

T1. Superior vena cava

T2. Brachiocephalic v.

T3. Subclavian v.

Arm (AR)

AR1. Cephalic v.

AR2. Brachial v.

AR3. Basilic v.

Forearm (FO)

FO1. Ulnar v.

FO2. Radial v.

Abdomen (A)

A1. Hepatic v.

A2. Hepatic portal v.

A3. Gastric v.

A4. Splenic v.

A5. Inferior mesenteric v.

A6. Superior mesenteric v.

A7. Inferior vena cava

A8. Renal v.

A9. Gonadal v.
   (testicular v. in male, ovarian v. in female)

A10. Common iliac v.

A11. Internal iliac v.

A12. External iliac v.

Thigh (TH)

TH1. Femoral v.

TH2. Great saphenous v.

Leg (L)

L1. Popliteal v.

L2. Anterior tibial v.

L3. Posterior tibial v.

Major Veins

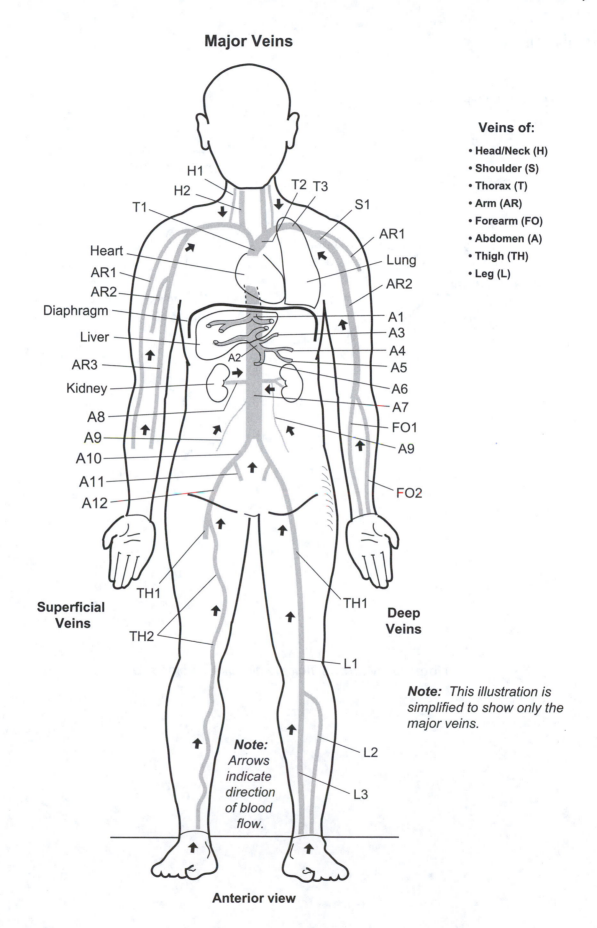

Veins of:

- **Head/Neck (H)**
- **Shoulder (S)**
- **Thorax (T)**
- **Arm (AR)**
- **Forearm (FO)**
- **Abdomen (A)**
- **Thigh (TH)**
- **Leg (L)**

H1
H2
T1
Heart
AR1
AR2
Diaphragm
Liver
AR3
Kidney
A8
A9
A10
A11
A12

T2 T3
S1
AR1
Lung
AR2
A1
A3
A4
A5
A6
A7
FO1
A9
FO2

A2

Superficial Veins

TH1
TH2

TH1

Deep Veins

L1

L2

L3

Note: Arrows indicate direction of blood flow.

Note: *This illustration is simplified to show only the major veins.*

Anterior view

Key to Illustration

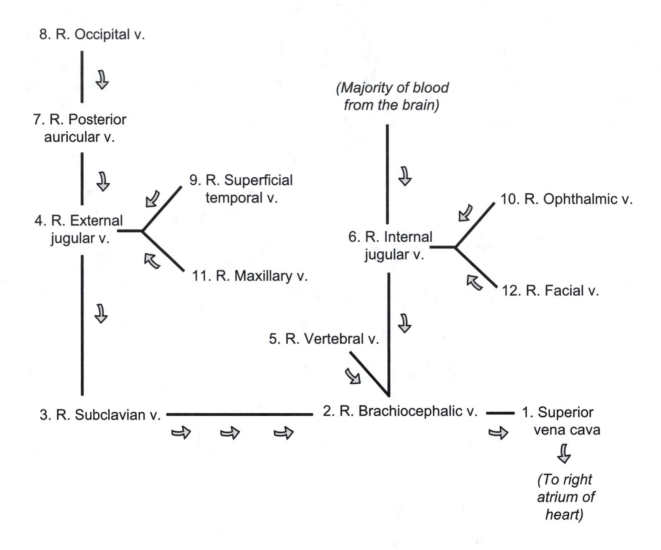

8. R. Occipital v.

7. R. Posterior
   auricular v.

(Majority of blood
from the brain)

9. R. Superficial
   temporal v.

4. R. External
   jugular v.

10. R. Ophthalmic v.

6. R. Internal
   jugular v.

11. R. Maxillary v.

12. R. Facial v.

5. R. Vertebral v.

3. R. Subclavian v.

2. R. Brachiocephalic v.

1. Superior
   vena cava

(To right
atrium of
heart)

Note: Gray arrows indicate direction of blood flow.

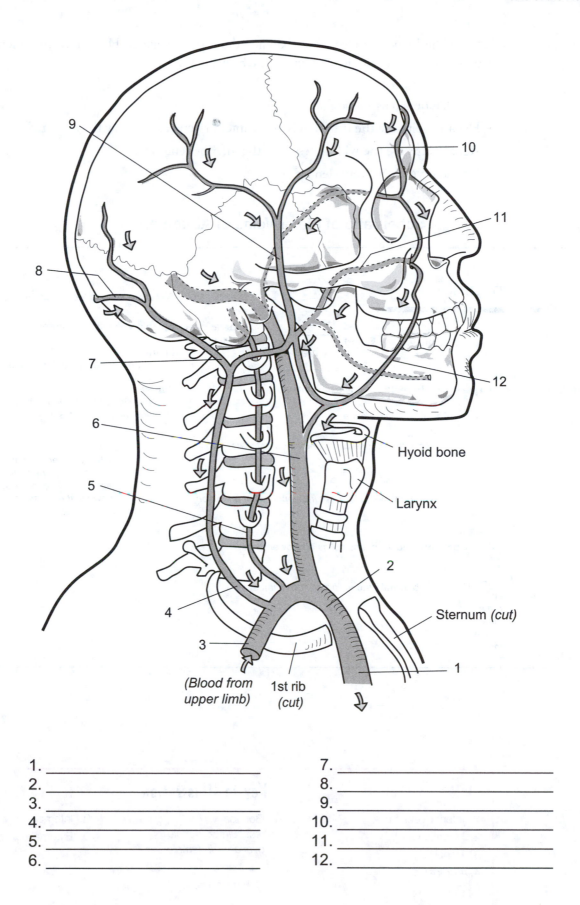

Hyoid bone

Larynx

Sternum *(cut)*

(Blood from upper limb)

1st rib *(cut)*

1. _____   7. _____
2. _____   8. _____
3. _____   9. _____
4. _____   10. _____
5. _____   11. _____
6. _____   12. _____

Description

Most of the blood vessels in the abdomen drain deoxygenated blood into the inferior vena cava, which empties into the right atrium of the heart.

Study Tips

- Use diaphragm as a landmark
- Hepatic veins are the first vessels of the inferior vena cava below the diaphragm
- **Renal veins** are the **widest** vessels of the inferior vena cava
- **Gonadal veins** are very **slender**

Schematic of Major Veins of Abdomen

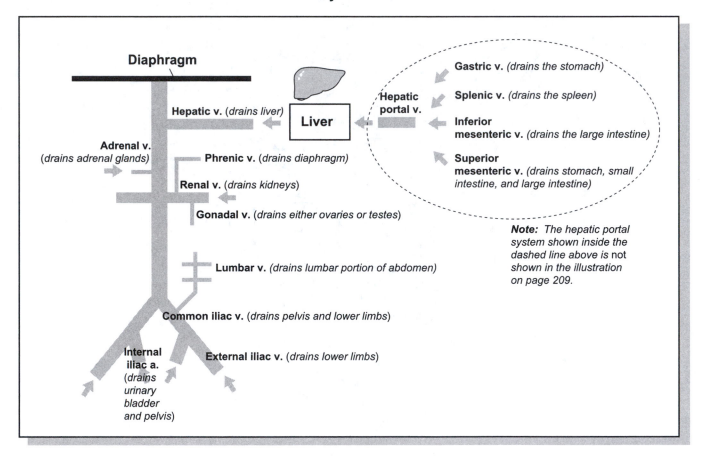

Diaphragm

Hepatic v. (drains liver)

Adrenal v. (drains adrenal glands)

Phrenic v. (drains diaphragm)

Renal v. (drains kidneys)

Gonadal v. (drains either ovaries or testes)

Lumbar v. (drains lumbar portion of abdomen)

Common iliac v. (drains pelvis and lower limbs)

Internal iliac a. (drains urinary bladder and pelvis)

External iliac v. (drains lower limbs)

Liver

Hepatic portal v.

Gastric v. (drains the stomach)

Splenic v. (drains the spleen)

Inferior mesenteric v. (drains the large intestine)

Superior mesenteric v. (drains stomach, small intestine, and large intestine)

Note: The hepatic portal system shown inside the dashed line above is **not** shown in the illustration on page 209.

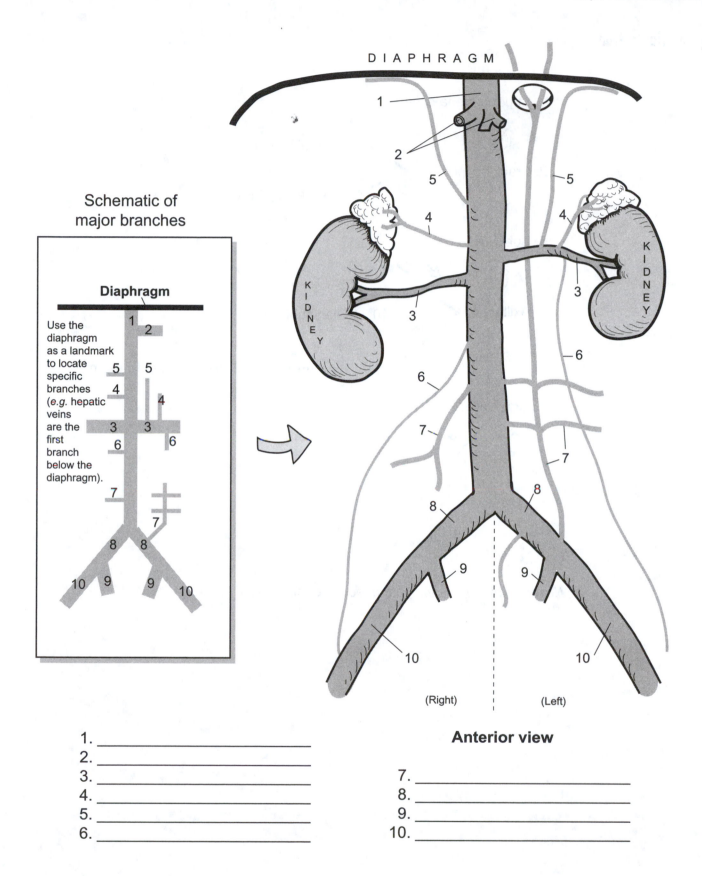

Schematic of major branches

Diaphragm

Use the diaphragm as a landmark to locate specific branches (*e.g.* hepatic veins are the first branch below the diaphragm).

DIAPHRAGM

KIDNEY

KIDNEY

(Right)   (Left)

Anterior view

1. _____
2. _____
3. _____
4. _____
5. _____
6. _____

7. _____
8. _____
9. _____
10. _____

Key to Illustration

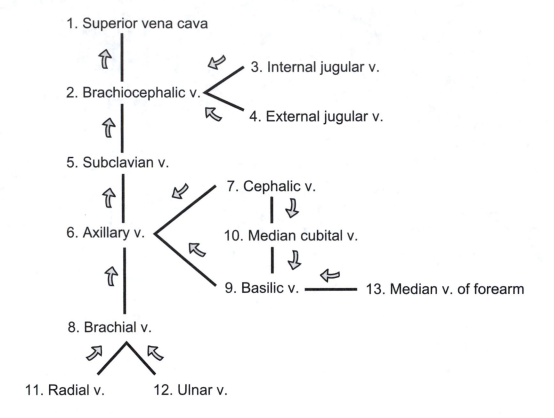

1. Superior vena cava
2. Brachiocephalic v.
3. Internal jugular v.
4. External jugular v.
5. Subclavian v.
6. Axillary v.
7. Cephalic v.
10. Median cubital v.
9. Basilic v.
13. Median v. of forearm
8. Brachial v.
11. Radial v.
12. Ulnar v.

Note: Gray arrows indicate direction of blood flow.

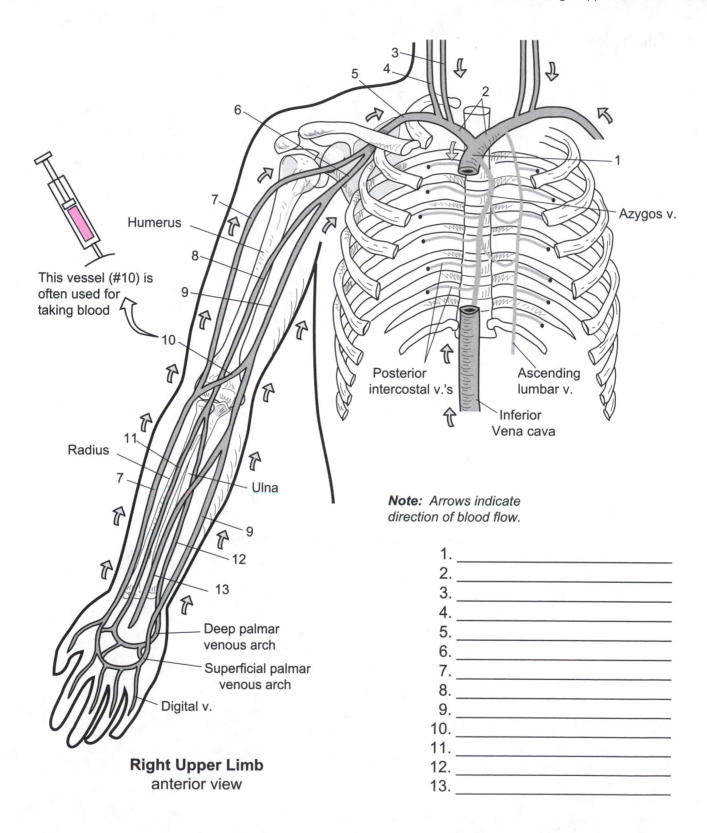

3

5

4

6

2

1

Azygos v.

Humerus

This vessel (#10) is often used for taking blood

7

8

9

10

Posterior intercostal v.'s

Ascending lumbar v.

Inferior Vena cava

Radius

11

7

Ulna

9

12

13

Deep palmar venous arch

Superficial palmar venous arch

Digital v.

Right Upper Limb
anterior view

Note: *Arrows indicate direction of blood flow.*

1. _____
2. _____
3. _____
4. _____
5. _____
6. _____
7. _____
8. _____
9. _____
10. _____
11. _____
12. _____
13. _____

Key to Illustration

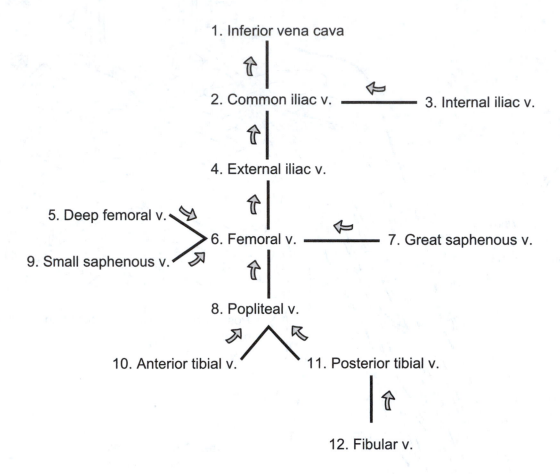

1. Inferior vena cava
2. Common iliac v.
3. Internal iliac v.
4. External iliac v.
5. Deep femoral v.
6. Femoral v.
7. Great saphenous v.
8. Popliteal v.
9. Small saphenous v.
10. Anterior tibial v.
11. Posterior tibial v.
12. Fibular v.

Note: Gray arrows indicate direction of blood flow.

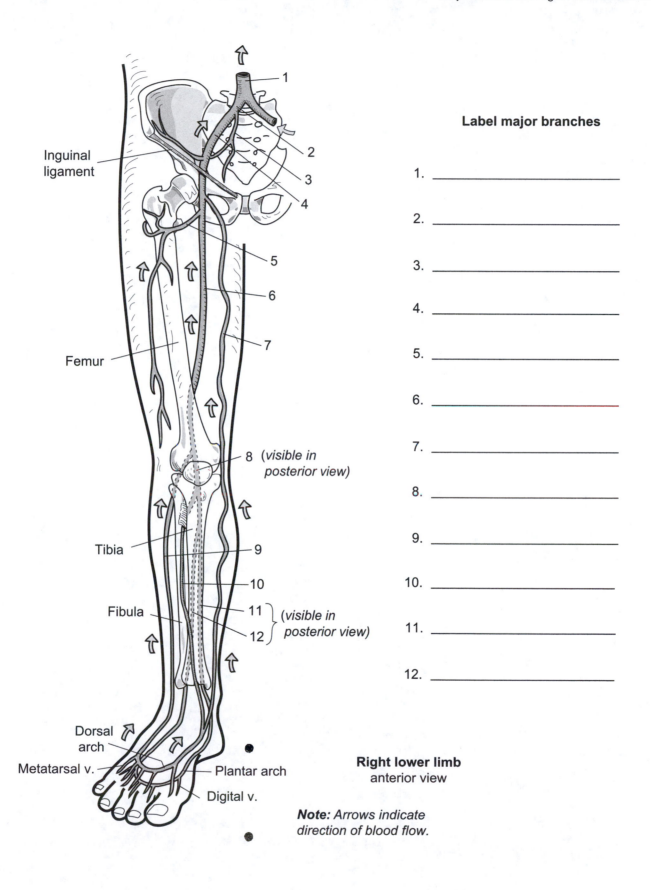

Inguinal ligament

Femur

Tibia

Fibula

Dorsal arch

Metatarsal v.

Plantar arch

Digital v.

8 (*visible in posterior view*)

11 } (*visible in posterior view*)
12 }

Right lower limb
anterior view

Note: *Arrows indicate direction of blood flow.*

Label major branches

1. _____

2. _____

3. _____

4. _____

5. _____

6. _____

7. _____

8. _____

9. _____

10. _____

11. _____

12. _____

Notes

Lymphatic System

Description

The cardiovascular system has a close relationship with the lymphatic system. Like the veins running through the body, the lymphatic system consists of a network of thin-walled vessels called **lymphatic vessels**. Like veins, they contain one-way valves (semilunar valves) that assist in circulating the lymph, which is under very low pressure.

Instead of carrying blood, the lymphatic vessels carry **lymph**—tissue fluid that was filtered from the blood. In the illustration, the gray areas indicate this filtration process. The composition of lymph is similar to plasma—mostly water along with some solutes such as salts.

Lymphatic vessels are connected to lymphatic capillaries and lymph nodes. **Lymphatic capillaries** are structurally similar to blood capillaries. Both are microscopic networks made of a single layer of simple squamous epithelium, but lymphatic capillaries contain flap-like structures that make them more permeable than blood capillaries. **Lymph nodes** are pea-sized structures that act as tiny filters to clean the lymph.

Like an oil filter cleans the motor oil in your car's engine, the lymph nodes filter the debris out of your lymph. These nodes contain macrophages that ingest and destroy pathogens such as bacteria. Lymph enters a lymph node through an **afferent lymphatic vessel** and leaves through an **efferent lymphatic vessel**. In short, "unclean lymph in, clean lymph out." The cleansed lymph is returned to the cardiovascular system via the subclavian veins.

Flow of Lymph

Here is a summary of the flow of lymph through the lymphatic system:

Lymphatic capillaries ⟶ afferent lymphatic vessel ⟶ lymph node ⟶ efferent lymphatic vessel ⟶ subclavian veins

Lymph is really nothing more than filtered blood plasma. All blood capillaries constantly filter the blood as a result of the force of blood pressure (see p. 68). This fluid, called **interstitial fluid**, fills the interstitial spaces between body cells, bathing them in fluid. Filtration occurs at the higher-pressure arteriole end of the capillary. Although some of this fluid is reabsorbed at the venule end, there is still an excess amount. As interstitial fluid pressure builds in the interstitial spaces, it is shunted into the nearby lymphatic capillaries, which act as a drain for the excess fluid. Although this fluid has not changed its chemical composition in any way, once inside the lymphatic capillary, it now is called **lymph**. Think of this **fluid cycle** as recycling of our plasma. This helps maintain normal fluid levels in the blood. As shown in the illustration: **Plasma** is filtered to become **interstitial fluid**, which becomes **lymph**, which becomes plasma once again. The cycle is complete!

Study Tip

The terms *afferent* and *efferent* apply to multiple organ systems. Here is a way to distinguish them: Afferent as in Approach; Efferent as in Exit.

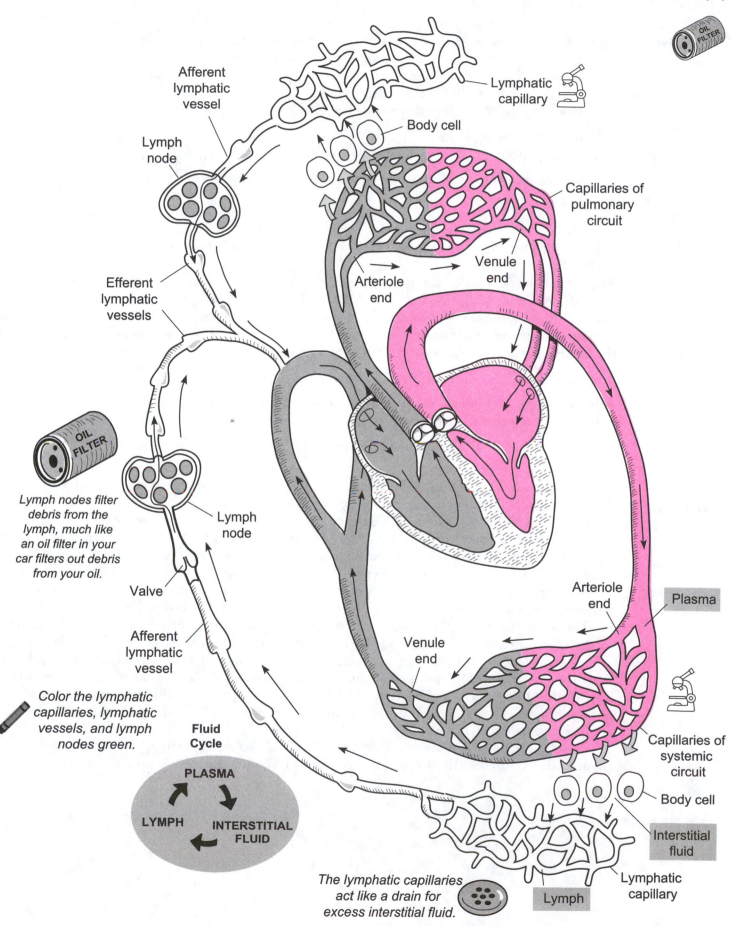

Afferent lymphatic vessel

Lymph node

Efferent lymphatic vessels

Lymphatic capillary

Body cell

Capillaries of pulmonary circuit

Venule end

Arteriole end

OIL FILTER

Lymph nodes filter debris from the lymph, much like an oil filter in your car filters out debris from your oil.

Lymph node

Valve

Afferent lymphatic vessel

Color the lymphatic capillaries, lymphatic vessels, and lymph nodes green.

Fluid Cycle

PLASMA

LYMPH — INTERSTITIAL FLUID

The lymphatic capillaries act like a drain for excess interstitial fluid.

Arteriole end

Plasma

Venule end

Capillaries of systemic circuit

Body cell

Interstitial fluid

Lymph

Lymphatic capillary

Description

This overview of the immune system has a focus on specific resistance. The **immune system** protects your body against foreign pathogens such as bacteria and viruses, using several lines of defense against these invaders. Think of a pathogenic invasion as an army of invaders attacking another army inside a medieval castle. The first line of defense is the wall around the castle. Similarly, your body has the following physical and chemical barriers as its first line of defense:

1. Physical barriers

   a. Skin: Thick layer of dead cells in the epidermis provides protection.

   b. Mucous membranes: Mucous film on these membranes traps microbes.

2. Chemical barriers

   a. Lysozyme in tears is an antibacterial agent.

   b. Gastric juice in the stomach is highly acidic (pH 2–3), which destroys bacteria.

The second line of defense consists of methods of nonspecific resistance that destroy invaders in a generalized way without targeting specific individuals. For example, if archers along the top of the castle were to shoot arrows into the invading army or if boiling oil were poured on them, this would help kill clusters of enemy soldiers. Similarly, your body has some general defenses for microbes that pass through the first lines of defense. A few examples are the following:

- Phagocytic cells ingest and destroy all microbes that pass into body tissues.

- Inflammation is a normal body response to tissue damage and other stimuli that brings more white blood cells to the site of pathogenic invasion.

- Fever inhibits bacterial growth and increases the rate of tissue repair during an infection.

Specific Resistance

The third line of defense deals with specific resistance, illustrated on the facing page. Think of these defenses like guided missiles that go after a specific target. In the medieval castle, they might be specially trained soldiers who act as assassins to kill the enemy's general. In short, they have a specific mission. Your immune system has "assassin" cells that attack microbes. Unlike in a war in which soldiers in different armies are wearing different uniforms, your immune system has a more difficult time distinguishing its own tissues ("self") from foreign microbes ("non-self"). To make this distinction, it relies on detecting **antigens**, which are specific substances found in foreign microbes. Most are proteins that serve as the stimulus to produce an immune response. The term "antigen" is coined from "**ANTI**-body **GEN**erating substances."

The illustration shows the immune response to an antigen. Once the antigen is detected, a dual response is activated by two groups of specialized lymphocytes called **T cells** and **B cells**. These cells are able to communicate with each other through chemical signaling. T cells typically are activated first. After they are activated, they can either directly destroy the microbes or use chemical secretions to destroy them. At the same time, T cells stimulate B cells to divide, forming other cells that are able to produce **antibodies**. These Y-shaped proteins circulate though the bloodstream and bind to specific antigens, thereby attacking microbes.

Details of the mechanisms that T cells and B cells use to attack microbes are examined in subsequent modules. Together, the T cells and B cells provide specific resistance to specific antigens.

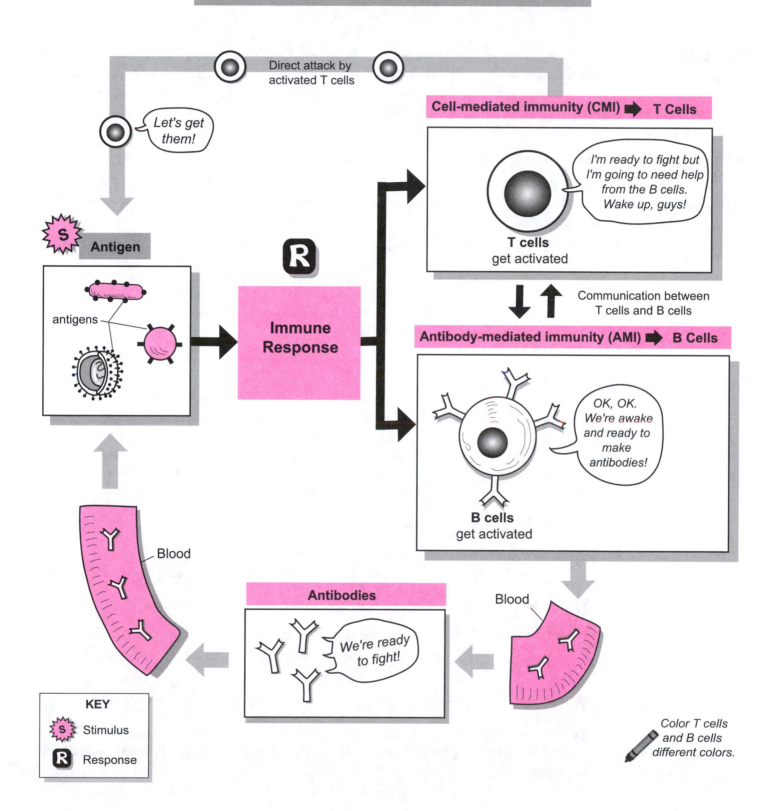

Immune System: Overview of Specific Resistance

Description

Cell-mediated immunity (CMI) involves the activation of **T cells** (T lymphocytes) by a specific antigen. In total, the body contains millions of different T cells—each able to respond to one specific antigen.

Development of T cells

T cells are a special type of lymphocyte. **Immature lymphocytes** are produced from stem cells in the **red bone marrow**. Some of these cells are processed within the **thymus gland**—hence, **T cell**—during embryological development, then released into the blood. These mature T cells are located in the blood, lymph, and lymphoid organs such as the lymph nodes and spleen.

Common T Cells and Their Functions

The three major types of T cells are as follows:

- **Cytotoxic T cells** secrete **lymphotoxin** and **perforin**. The former trigger destruction of the pathogen's DNA, and the latter create holes in the pathogen's plasma membrane, resulting in a **lysed cell**.

- **Helper T cells** secrete **interleukin 2** (I-2), which stimulates cell division of T cells and B cells. This can be thought of as recruiting more soldiers for the fight.

- **Memory T cells** remain dormant after the initial exposure to an antigen. If the same antigen presents itself again—even years later—the memory cells are stimulated to convert themselves into cytotoxic T cells and enter the fight.

Phagocytosis

Phagocytes ("eater cells") use the process of **phagocytosis** ("cell eating") to ingest foreign pathogens. An example is a macrophage ("big eater"), which is derived from the largest white blood cells—monocytes. Macrophages leave the bloodstream and enter body tissues to patrol for pathogens. As some phagocytic cells engage in phagocytosis, they present antigenic fragments on their plasma membrane surface, thereby stimulating the activation of T cells. Consequently, they are an important part of the CMI. A summary of the phagocytosis process shown in the illustration is:

1. **Microbe** attaches to **phagocyte.**

2. Phagocyte's **plasma membrane** forms arm-like extensions that surround and engulf the microbe. The encapsulated microbe pinches off from the plasma membrane to form a **vesicle**.

3. The vesicle merges with a **lysosome**, which contains **digestive enzymes**.

4. The digestive enzymes begin to break down the microbe. The phagocyte extracts the nutrients it can use, leaving the **indigestible material** and **antigenic fragments** within the vesicle.

5. The phagocyte makes **protein markers**, and they enter the vesicle.

6. The indigestible material is removed by exocytosis. The antigenic fragments bind to the protein marker and are displayed on the plasma membrane surface. This serves to activate T cells.

Development of T cells

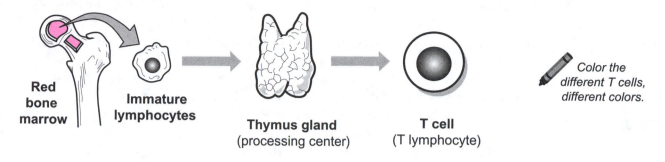

Red bone marrow → Immature lymphocytes → Thymus gland (processing center) → T cell (T lymphocyte)

Color the different T cells, different colors.

Common types of T cells and their functions

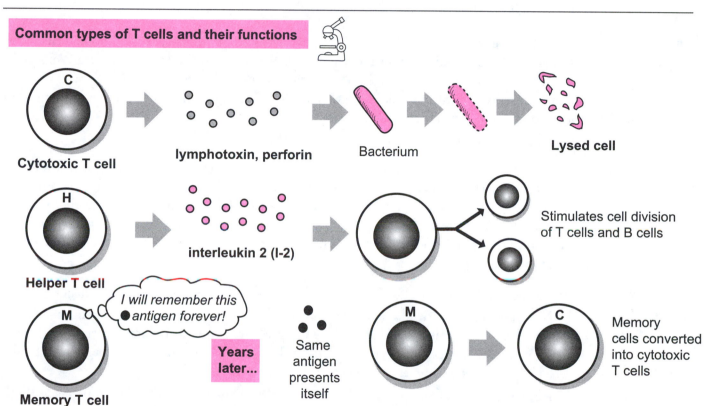

Cytotoxic T cell → lymphotoxin, perforin → Bacterium → → **Lysed cell**

Helper T cell → interleukin 2 (I-2) → Stimulates cell division of T cells and B cells

Memory T cell

I will remember this ● antigen forever!

Years later... Same antigen presents itself

M → C Memory cells converted into cytotoxic T cells

Phagocytosis

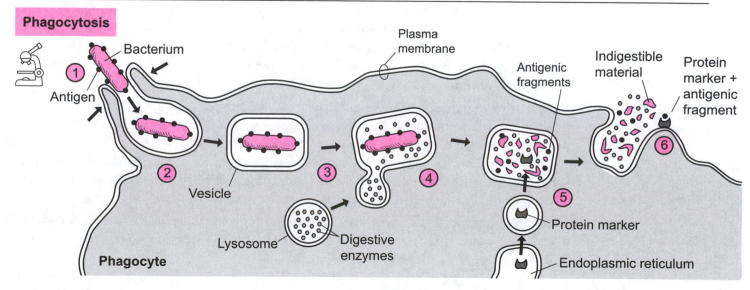

1 Bacterium, Antigen
2 Vesicle
Lysosome — Digestive enzymes
3
4 — Plasma membrane
5 — Protein marker
Endoplasmic reticulum
Antigenic fragments
6 Indigestible material — Protein marker + antigenic fragment

Phagocyte

Description

Antibody-mediated immunity (AMI) involves the activation of **B cells** (B lymphocytes) by a specific antigen. This triggers the B cells to transform into **plasma cells**, which are able to secrete special proteins called **antibodies**. The antibodies are transported through the blood and the lymph to the pathogenic invasion site. In total, the body contains millions of different B cells— each able to respond to one specific **antigen**. Amazing!

Development of B Cells

B cells are a special type of lymphocyte. **Immature B cells** are produced from stem cells in the **red bone marrow**. These immature cells are later processed within the red bone marrow—hence, B cell—during embryological development to become mature B cells, then are released into the blood. The mature B cells are located in the blood, lymph, and lymphoid organs such as the lymph nodes and spleen.

Antibody Production

B cells can be stimulated to divide, forming two types of numerous cells: (1) **plasma cells**, which secrete **antibodies,** and (2) **B memory cells** which exist in the body for many years, ensuring a quick response to the same antigen. Antibodies (immunoglobulin, or Ig) are Y-shaped proteins that are subdivided into five classes: IgG, IgM, IgA, IgE, IgD. These are listed in order from the *most* common to the *least* common.

Study Tip: Mnemonic: *Get Me Another Excellent Donut!*

The basic structure on an antibody consists of four polypeptide chains—two **heavy chains** and two **light chains**. Both heavy chains and both light chains are identical to the other, and each contains a **constant region** and a **variable region.** The constant region forms the "trunk" of the molecule, and the variable region forms the **antigen-binding site** on the antibody. Note that each antibody has two of these antigen-binding sites. Think of these like claws on a lobster used to "grab" its specific antigen.

How do Antibodies Work?

Antibodies work through many different mechanisms, of which the following are major ones:

1. **Neutralizing antigen:** the **antibody** can bind to an **antigen**, forming an **antigen–antibody complex**. This forms a shield around the antigen, preventing its normal function. In this way, a toxin from a bacterium may be neutralized or a viral antigen may not be able to bind to a body cell thereby preventing infection.

2. **Activating complement:** "complement" refers to a group of plasma proteins made by the liver that normally are inactive in the blood. An **antigen-antibody complex** triggers a cascade reaction that activates these proteins to induce beneficial responses. For example, some of these activated proteins can cluster together to form a pore or channel that inserts into a microbial plasma membrane. This results in a **lysed cell**. Other responses include **chemotaxis** and **inflammation**. Both of these mechanisms serve to increase the number of white blood cells at the site of invasion.

3. **Precipitating antigens:** numerous antibodies can bind to the same free antigens in solution to cross-link them. This cross-linked mass then precipitates out of solution, making it easier for phagocytic cells to ingest them by phagocytosis.

    Similarly, microbes (such as bacteria) can be clumped together by a process called *agglutination* (not illustrated). The antigens within the cell walls of the bacteria are what are cross-linked. As with precipitation, this is followed by phagocytosis.

4. **Facilitating phagocytosis:** an **antigen-antibody complex** acts like a warning sign to signal phagocytic cells to attack. In fact, the complex also binds to the surface of **macrophages** to further facilitate phagocytosis.

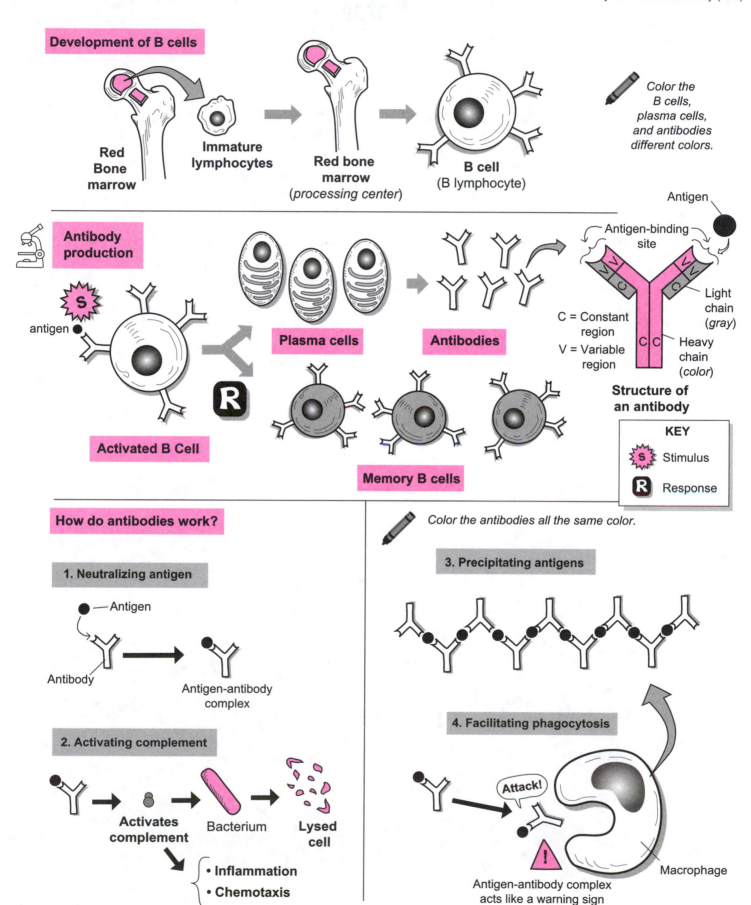

Development of B cells

Red Bone marrow → Immature lymphocytes → Red bone marrow (*processing center*) → B cell (B lymphocyte)

Color the B cells, plasma cells, and antibodies different colors.

Antibody production

antigen

S

Activated B Cell

R

Plasma cells

Antibodies

Memory B cells

Antigen

Antigen-binding site

C = Constant region
V = Variable region

Light chain (*gray*)

Heavy chain (*color*)

Structure of an antibody

KEY

S Stimulus

R Response

How do antibodies work?

1. Neutralizing antigen

Antigen

Antibody

Antigen-antibody complex

2. Activating complement

Activates complement → Bacterium → Lysed cell

• Inflammation
• Chemotaxis

Color the antibodies all the same color.

3. Precipitating antigens

4. Facilitating phagocytosis

Attack!

Macrophage

Antigen-antibody complex acts like a warning sign

223

Notes

Respiratory System

Description

The respiratory system is divided into two major divisions: *upper respiratory system* and *lower respiratory system*. The **upper respiratory system** consists of the nose, nasal cavity, paranasal sinuses, and pharynx. The **lower respiratory system** consists of the larynx, trachea, bronchi, and lungs.

Let's trace the pathway of a molecule of oxygen (O_2) through the respiratory system to its final destination at a body cell. The O_2 molecule enters the **nasal cavity** through the **external nares** (*nostrils*). As it passes to the back of this moist chamber, it enters the **nasopharynx**, then the **oropharynx**, and finally the **laryngopharynx**.

After passing the rigid, flap-like structure called the **epiglottis**, it enters the **larynx**, then passes through the slit-like opening between the vocal cords called the **glottis**. Next it moves through the long, rigid tube of the **trachea** until it reaches a split in this passageway. Following the passageway branching into the left lung, it enters the **left primary bronchus**, then enters the next split in the passageway, the narrower **secondary bronchus**. The next branch is the even narrower **tertiary bronchus**. Finally, the O_2 molecule enters a microscopic tube called a **bronchiole**.

This continues to branch into a **terminal bronchiole**, then a **respiratory bronchiole**, and finally terminates in an air sac called an **alveolus**. This delicate air sac is the end of the bronchial tree in the lungs. The wall of each alveolus is made of simple squamous epithelium, which allows for easy diffusion of the O_2 molecule out of the alveolus and into the bloodstream, where it will be delivered to a body cell.

Analogies

- The **larynx** looks like the **head of a snapping turtle**. The **turtle's head** is the **thyroid cartilage** and the **turtle's lower jaw** is the **cricoid cartilage**. The **neck of the turtle** is the **trachea**.

- The **clusters of alveoli** are like a **wad of bubble wrap** used in packaging. Both are small, sac-like structures filled with air.

Study Tip

Palpate (*feel by touch*):

- On average, the larynx is slightly larger in males than in females, but you can easily feel a portion of it in either gender. The main structure you feel beneath the skin is the thyroid cartilage commonly called the *Adam's apple*. The portion of the Adam's apple that protrudes most anteriorly is called the laryngeal prominence.

Key to Illustration

Larynx (L)

L1. Thyrohyoid membrane
L2. Thyroid cartilage
L3. Laryngeal prominence
L4. Cricothyroid ligament
L5. Cricoid cartilage

Upper Respiratory Tract (UP)

UP1. Frontal sinus
UP2. Sphenoidal sinus
UP3. External nares (nostrils)
UP4. Nasal cavity
UP5. Superior concha
UP6. Middle concha
UP7. Inferior concha
UP8. Pharyngeal tonsil
UP9. Nasopharynx
UP10. Palatine tonsil
UP11. Oropharynx
UP12. Lingual tonsil
UP13. Epiglottis
UP14. Laryngopharynx

Lower Respiratory Tract (LO)

LO1. Vocal fold
LO2. Trachea
LO3. Tracheal rings
LO4. Parietal pleura
LO5. Visceral pleura
LO6. Location of carina (*internal ridge*)
LO7. Primary bronchus
LO8. Secondary bronchus
LO9. Tertiary bronchus
LO10. Terminal bronchiole
LO11. Respiratory bronchiole
LO12. Alveoli
LO13. Simple squamous epithelium

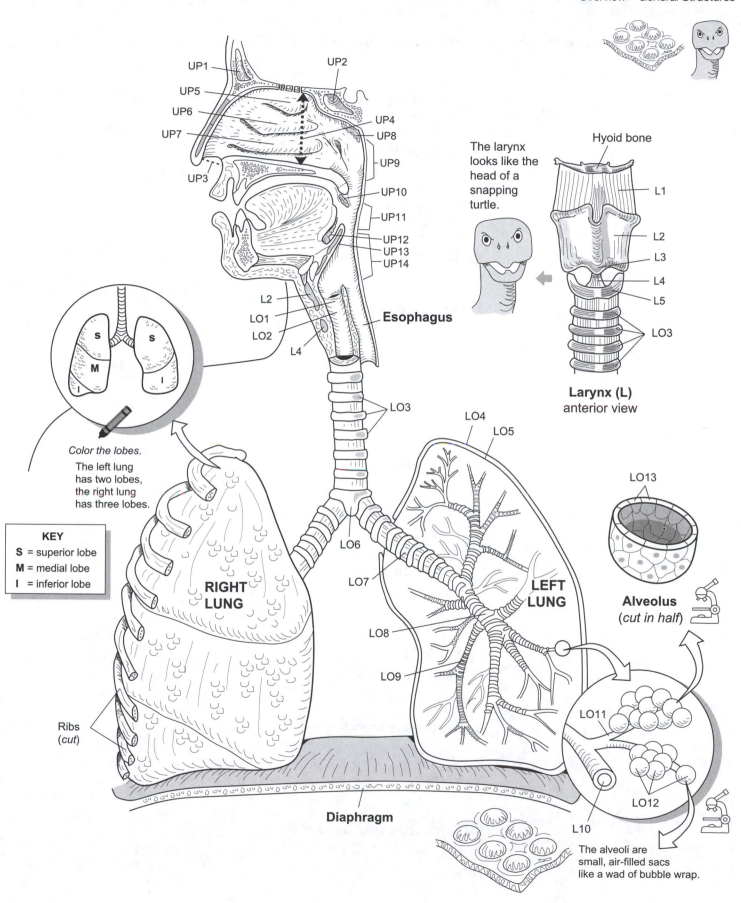

UP1
UP5
UP6
UP7
UP3
UP2
UP4
UP8
UP9
UP10
UP11
UP12
UP13
UP14

L2
LO1
LO2
L4

Esophagus

The larynx looks like the head of a snapping turtle.

Hyoid bone

L1
L2
L3
L4
L5

LO3

Larynx (L)
anterior view

Color the lobes.
The left lung has two lobes, the right lung has three lobes.

KEY
S = superior lobe
M = medial lobe
I = inferior lobe

RIGHT LUNG

Ribs
(*cut*)

LO3

LO6

LO7

LO8

LO9

LO4
LO5

LEFT LUNG

LO13

Alveolus
(*cut in half*)

LO11

LO12

L10

The alveoli are small, air-filled sacs like a wad of bubble wrap.

Diaphragm

227

The following flowchart gives the pathway of oxygen through the respiratory system and into a body cell:

1. External nares *(nostrils)*

2. Nasal cavity

3. Nasopharynx

4. Oropharynx

5. Laryngopharynx

6. Larynx

7. Trachea

8. Primary bronchus

9. Secondary bronchus

10. Tertiary bronchus

11. Bronchiole

12. Terminal bronchiole

13. Respiratory bronchiole

14. Alveolus

15. Erythrocyte *(red blood cell)*

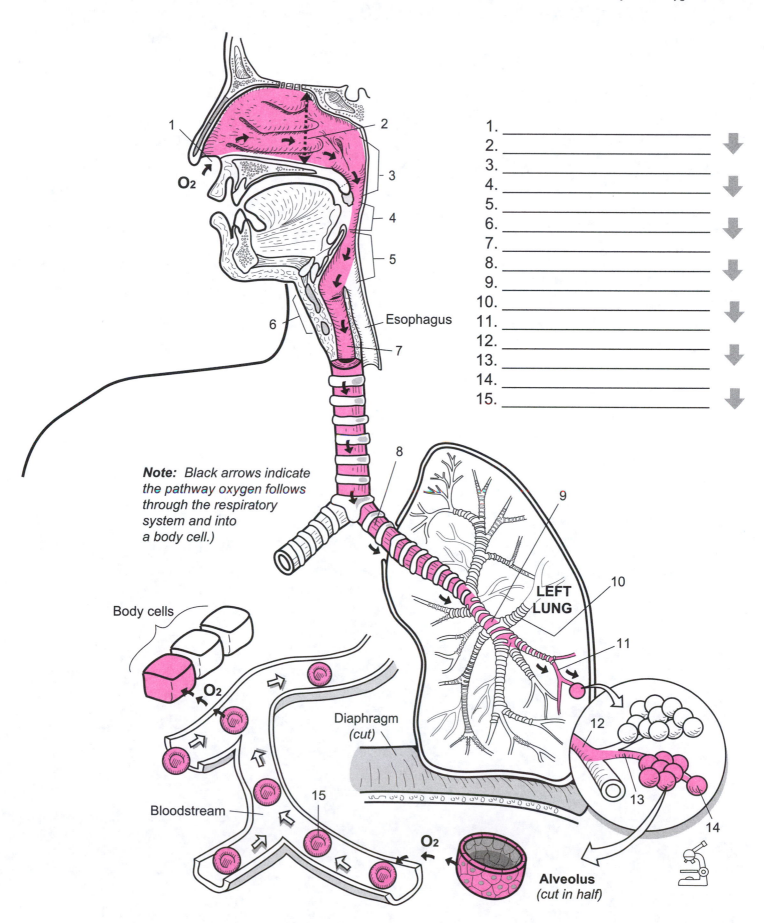

O₂

1 _____
2 _____

3 _____

4 _____

5 _____

Esophagus

6 _____

7 _____

1. _____
2. _____
3. _____
4. _____
5. _____
6. _____
7. _____
8. _____
9. _____
10. _____
11. _____
12. _____
13. _____
14. _____
15. _____

Note: *Black arrows indicate the pathway oxygen follows through the respiratory system and into a body cell.)*

8 _____

9 _____

LEFT LUNG

10 _____

11 _____

Body cells

O₂

Diaphragm *(cut)*

12 _____

13 _____

14 _____

Bloodstream

15 _____

O₂

Alveolus *(cut in half)*

229

Notes

Digestive System

Description

Let's follow a **bolus** (*compressed mass*) of food through the digestive tract. At lunchtime you are eating a chicken sandwich that contains many nutrients the body needs. For example, to name just a few, there is **protein** in the chicken, **complex carbohydrates** (*starch*) in the bread, and **lipids** (*fats*) in the mayonnaise.

You take a bite of the sandwich and it enters the **oral cavity**. Three different pairs of **salivary glands** (*parotid, sublingual,* and *submandibular*) release saliva into the cavity. The saliva lubricates the bolus to aid in swallowing. Within the saliva is an enzyme called **amylase** that assists in chemical digestion of the complex carbohydrates in the bread of the sandwich. After mechanically breaking down the food by chewing, you begin to swallow the bolus. Assisted by numerous muscles, it moves into the **oropharynx** and slides against the **epiglottis**, which folds back to cover the opening to the trachea and divert the bolus into the esophagus. As it moves through the esophagus by rhythmic waves of smooth muscle contraction called **peristalsis**, it gets lubricated again by **mucus**. Now it passes through the lower esophageal sphincter and into the **stomach**. Here the acidic gastric secretions destroy bacteria in the bolus. An enzyme made by the stomach called **pepsin** begins digestion of the protein from the chicken. The mixture of food and gastric secretions from the stomach is now called **chyme**. The chyme moves through the **pyloric sphincter** and begins the 20-ft. long journey through the small intestine. From beginning to end, this tube consists of the **duodenum, jejunum,** and **ileum**. In the duodenum, pancreatic enzymes are added to the chyme and **bile** from the gallbladder. Bile and pancreatic **lipase** help digest lipids. In the small intestine, the protein from the chicken, complex carbohydrate from the bread, and lipid from the mayonnaise will finally be chemically broken down into their smallest usable components. Protein is broken down into **amino acids**, and complex carbohydrates into simple carbohydrates such as **glucose,** and **lipids** (*fats*) into **fatty acids** and **glycerol**. These final products will then be absorbed into the bloodstream and transported to cells in the body to be used as nutrients.

The remaining nutrients and waste products move out of the small intestine through the **ileocecal valve** and into the 5 ft. **large intestine** (*large bowel*). Most nutrients have already been absorbed in the small intestine but the large intestine still needs to reabsorb water and electrolytes into the blood. Finally, indigestible waste products move through the **rectum** and out the **anus** for elimination from the body.

Key to Illustration

1. Oral cavity
2. Sublingual salivary gland
3. Submandibular salivary gland
4. Parotid salivary gland
5. Tongue
6. Epiglottis
7. Esophagus
8. Stomach

Small Intestine

9. Duodenum
10. Jejunum
11. Ileum

Large Intestine

12. Cecum
13. Ascending colon
14. Transverse colon
15. Descending colon
16. Sigmoid colon
17. Rectum
18. Anus

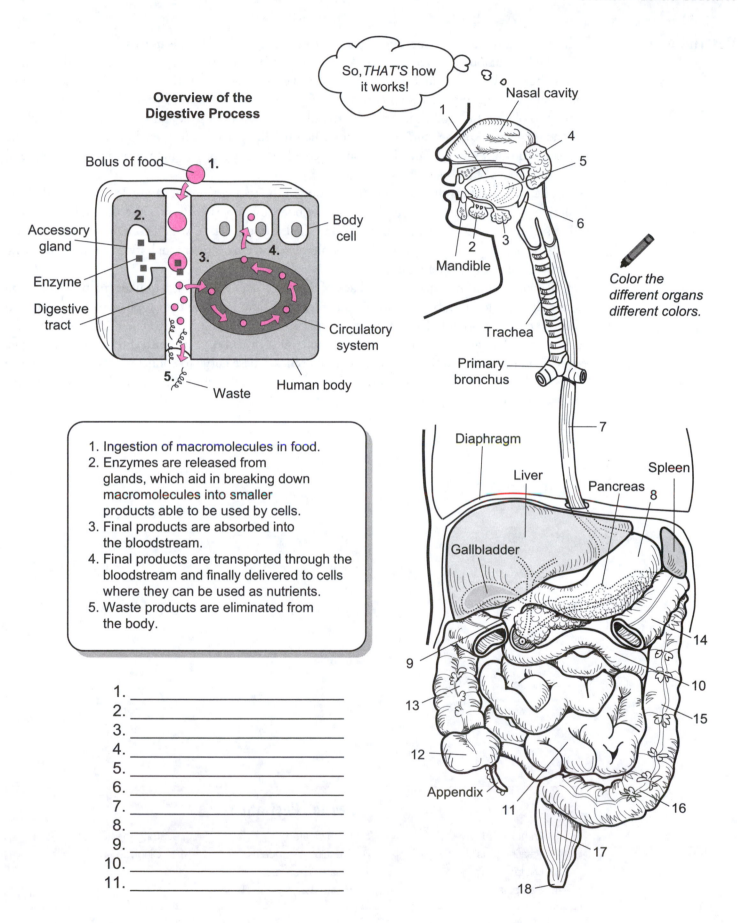

Overview of the Digestive Process

Bolus of food

1.

2. Accessory gland

Enzyme

Digestive tract

3.

4.

Body cell

Circulatory system

Human body

5. Waste

So, *THAT'S* how it works!

Nasal cavity

1

4

5

6

2 Mandible

3

Trachea

Primary bronchus

7

Color the different organs different colors.

1. Ingestion of macromolecules in food.
2. Enzymes are released from glands, which aid in breaking down macromolecules into smaller products able to be used by cells.
3. Final products are absorbed into the bloodstream.
4. Final products are transported through the bloodstream and finally delivered to cells where they can be used as nutrients.
5. Waste products are eliminated from the body.

Diaphragm

Liver

Spleen

Pancreas

8

Gallbladder

9

13

12

Appendix

11

14

10

15

16

17

18

1. _____
2. _____
3. _____
4. _____
5. _____
6. _____
7. _____
8. _____
9. _____
10. _____
11. _____

Description

The innermost lining of the **buccal cavity** (*oral cavity*) is called the **mucosa**. It is composed of **stratified squamous epithelium** (non-keratinized), which coats the roof and floor of the mouth, lines the inside of the cheeks, and forms a ridge of tissue below the teeth called the gums or **gingivae**. The mucosa curves anteriorly to form the upper and lower lips, or **labia**. Two separate, flattened plates of mucosa, each called a **frenulum,** anchor the upper and lower lips to the midline of the gingivae. Another similar structure called the **lingual frenulum** anchors the tongue to the floor of the oral cavity. The thick **tongue** covers the floor of the mouth and contains several different groups of muscles.

The **hard palate** is the most anterior portion of the roof of the oral cavity. It is composed of the palatine process of the maxilla and the palatine bone and functions to separate the oral cavity from the nasal cavity. Behind it lies a fleshy plate called the **soft palate**, which contains no bone within it. It separates the oral cavity from the nasopharynx and covers the nasopharynx during swallowing to prevent food from entering it.

Dangling from the middle of the back of the soft palate is a small, flap-like structure called the **uvula**. It holds food in the oral cavity and prevents it from entering the **oropharynx** too soon. Two sets of arches are found at the back of the oral cavity. From the front to the back, the first set is called the **palatoglossal arches**. These curve to connect the soft palate with the base of the tongue. The second set is called the **palatopharyngeal arches**. These curve to connect the soft palate to the sides of the pharynx. Between these arches, on either side of the mouth, are two masses of lymphatic tissue called the **palatine tonsils**.

Key to Illustration

1. Frenulum of lower lip	5. Hard palate	9. Tongue
2. Palatine tonsil	6. Frenulum of upper lip	10. Gingivae
3. Uvula	7. Palatoglossal arch	
4. Soft palate	8. Palatopharyngeal arch	

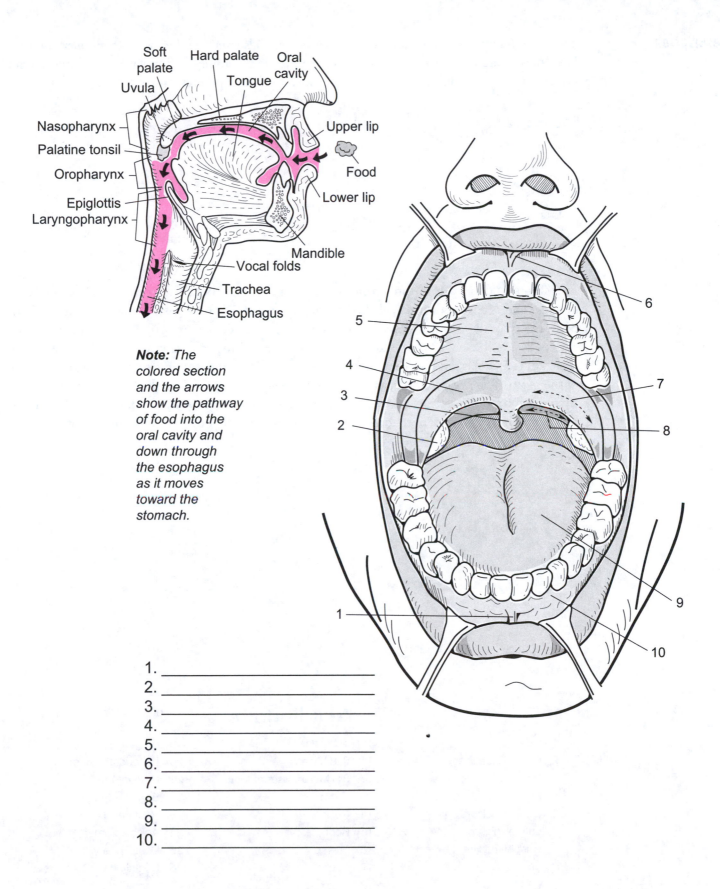

Soft palate

Hard palate

Uvula

Oral cavity

Tongue

Nasopharynx

Upper lip

Palatine tonsil

Oropharynx

Food

Epiglottis

Lower lip

Laryngopharynx

Mandible

Vocal folds

Trachea

Esophagus

Note: The colored section and the arrows show the pathway of food into the oral cavity and down through the esophagus as it moves toward the stomach.

1. _____
2. _____
3. _____
4. _____
5. _____
6. _____
7. _____
8. _____
9. _____
10. _____

Description

There are four different types of teeth—incisors, cuspids (*canines*), bicuspids (*premolars*), and molars. Incisors are for cutting, cuspids are for tearing, and bicuspids and molars are for crushing and grinding food. The adult jaws can accommodate a total of 32 permanent teeth.

Tooth Type	Upper Jaw	Lower Jaw	Total
Incisor	4	4	8
Cuspid	2	2	4
Bicuspid	4	4	8
Molar	6	6	12
		Total	32

Each tooth is divided into three regions: **crown, neck,** and **root**. The crown is visible above the gumline and is covered with a calcified coating of **enamel**—the hardest substance produced in the body. Deep to this covering is a bone-like substance called **dentin**, which makes up the majority of the tooth. At the center of the tooth is a chamber called the **pulp cavity**, which contains spongy tissue, blood vessels, and nerves. A bony substance called **cementum** covers the dentin in the root of the tooth. The **periodontal ligaments** anchor the root of the tooth to the bone in the jaw. At the tip of each root is an opening called the **apical foramen** that allows the **dental artery, dental vein,** and **dental nerve**, to penetrate through the narrow **root canal** and up into the **pulp cavity**.

Key to Illustration

1. Incisors
2. Cuspids (*canines*)
3. Bicuspids (*premolars*)
4. Molars
5. Central incisors
6. Lateral incisor
7. Cuspid (*canine*)
8. 1st Premolar
9. 2nd Premolar
10. 1st Molar
11. 2nd Molar
12. 3rd Molar (*wisdom tooth*)
13. Crown
14. Neck
15. Root
16. Enamel
17. Dentin
18. Pulp cavity
19. Gingiva
20. Periodontal ligaments
21. Cementum
22. Bone
23. Root canal
24. Apical foramen
25. Dental artery
26. Dental nerve
27. Dental vein

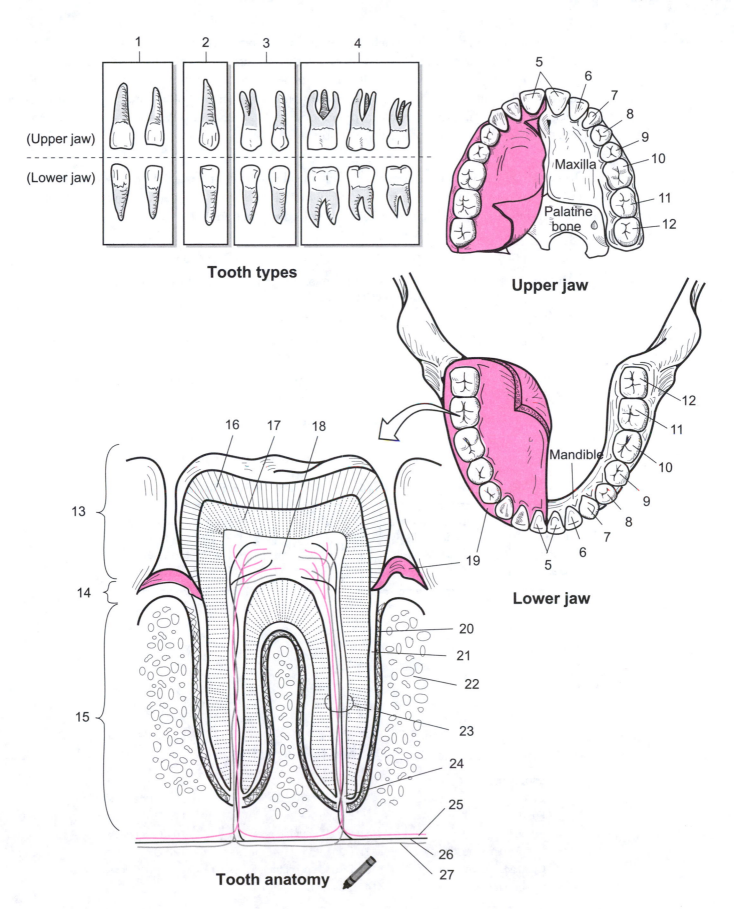

Tooth types

(Upper jaw)

(Lower jaw)

1 2 3 4

Upper jaw

5 6 7 8 9 10 11 12

Maxilla

Palatine bone

Lower jaw

12 11 10 9 8 7 6 5

Mandible

19

Tooth anatomy

13 14 15 16 17 18 19 20 21 22 23 24 25 26 27

Description

The stomach connects the esophagus to the first portion of the small intestine, called the duodenum. It is a specialized sac that contains three layers of smooth muscle within its wall. The innermost lining of the stomach, called the **mucosa**, is coated with a protective layer of alkaline mucus. Folds of mucosa called **rugae** allow the stomach to increase surface area to maximize **gastric juice**. This acidic mixture contains hydrochloric acid and enzymes. When this gastric juice combines with food, a mixture called **chyme** is formed.

Function

To release gastric juice for the purpose of chemical digestion of food, the mucosa manufactures an enzyme called **pepsin**, which initiates protein digestion. Within the mucosal lining of the stomach are four different types of cells that have the following functions:

Mucosal Cell	Function
Mucous cells	Secrete an alkaline **mucus** to protect the stomach lining from acidic gastric juice
Chief cells	Secrete the inactive enzyme—**pepsinogen**
Parietal cells	Secrete hydrochloric acid (HCl); this helps convert pepsinogen into the active enzyme **pepsin**
G cells	Produce and secrete the hormone **gastrin**, which increases secretion from parietal and chief cells. It also induces smooth muscle contraction in the stomach wall.

Study Tip

Use the following mnemonics to recall the mucosal cell types and the substances they produce:

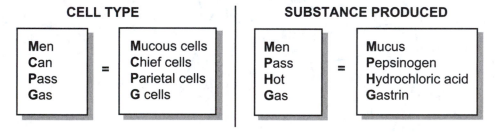

CELL TYPE

Men		Mucous cells
Can	=	Chief cells
Pass		Parietal cells
Gas		G cells

SUBSTANCE PRODUCED

Men		Mucus
Pass	=	Pepsinogen
Hot		Hydrochloric acid
Gas		Gastrin

Key to Illustration

1. Esophagus
2. Cardia
3. Fundus
4. Serosa
5. Longitudinal muscle layer
6. Circular muscle layer
7. Oblique muscle layer
8. Mucosa
9. Lesser curvature
10. Body
11. Greater curvature
12. Rugae
13. Pyloric sphincter
14. Gastric pit
15. Gastric gland
16. Mucous cells
17. Parietal cells
18. Chief cells
19. G cells
20. Muscularis mucosa

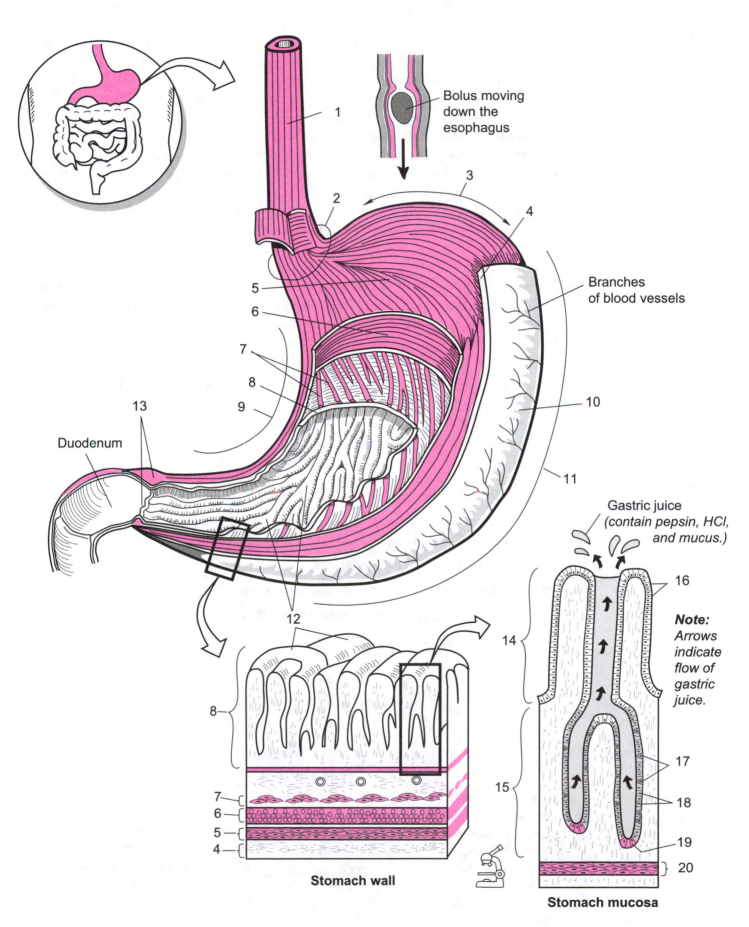

Bolus moving
down the
esophagus

Branches
of blood vessels

Duodenum

Gastric juice
*(contain pepsin, HCl,
and mucus.)*

Note:
*Arrows
indicate
flow of
gastric
juice.*

Stomach wall

Stomach mucosa

239

Description

The **small intestine** is a hollow, muscular tube approximately 20 feet in length that links the stomach to the large intestine. It is subdivided into three parts—**duodenum** (10 in.), **jejunum** (8 ft.), and **ileum** (12 ft.). The jejunum and ileum are loosely held in place by a highly vascular serous membrane called **mesentery**, which anchors the small intestine to the posterior wall of the abdominal cavity. From outermost to innermost, the layers in the wall of the small intestine are the **serosa**, **muscularis externa**, **submucosa**, and **mucosa**.

Folds are used to increase surface area anywhere in the body. Within the small intestine, there are three significant folded structures: **plicae circulares**, **villi**, and **microvilli**. All of these are microscopic except the plica circulares. These folds increase the total surface area to facilitate the absorption of nutrients. The **villi** are finger-like projections that extend from the plica and are surrounded by a layer of **simple columnar epithelium**. **Goblet cells** scattered within this tissue secrete mucus.

Each simple columnar epithelial cell has folds in its plasma membrane called microvilli. Nutrients must pass through the microvilli, then through the cell, in order to enter the villus. Within the villus is a **blood capillary** and a specialized lymphatic capillary called a **lacteal**. Depending on the type of nutrient, once inside the villus, it will enter either the blood capillary or the lacteal to complete its absorption.

Analogy

The **folds** on the inside of the **small intestine** are like a **folded carpet sample**. The **fold** is the **plica circulares**. The **carpet fibers** sticking out from this sample are the **villi**.

Location

Abdominal cavity; surrounded by the large intestine; below the stomach

Function

Absorption of nutrients

- **water, vitamins, minerals**
- **amino acids** (*from the digestion of proteins*)
- **fatty acids, glycerol** (*from the digestion of lipids*)
- **monosaccharides** (*from the digestion of complex carbohydrates*)
- **nucleotides** (*from the digestion of nucleic acids such as DNA*)

Key to Illustration

Wall of Small Intestine
1. Serosa
2. Muscularis externa (*longitudinal layer*)
3. Muscularis externa (*circular layer*)

4. Submucosa
5. Mucosa

Villus Structures
6. Arteriole
7. Blood capillary
8. Lacteal

9. Simple columnar epithelial cell
10. Goblet cell
11. Venule
12. Plasma (*cell*) membrane
13. Nucleus of one cell

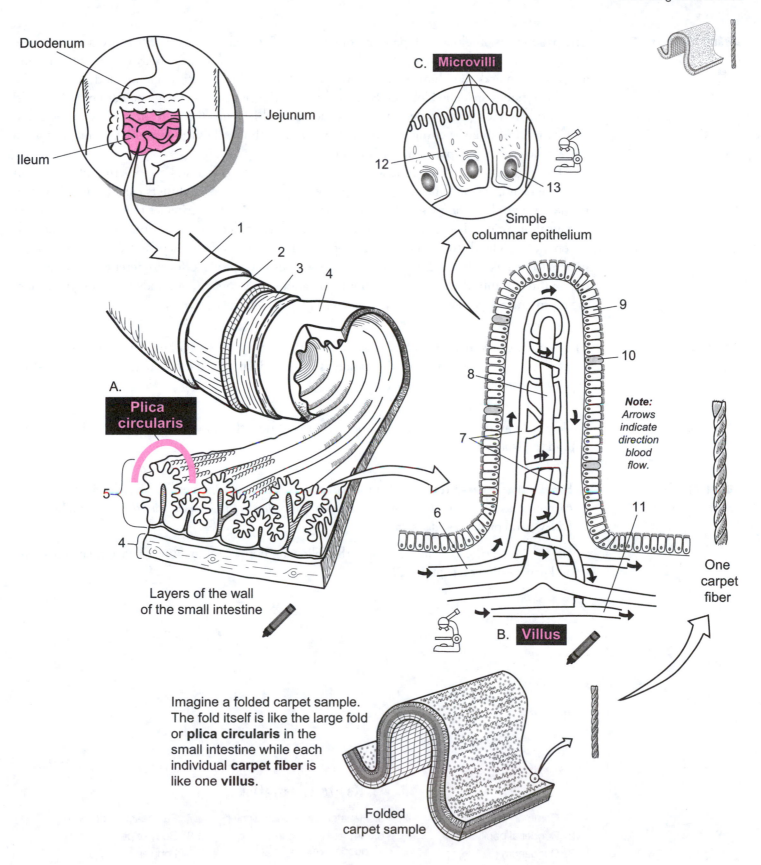

Duodenum

Jejunum

Ileum

C. **Microvilli**

12

13

Simple
columnar epithelium

1

2

3

4

A.

**Plica
circularis**

5

4

Layers of the wall
of the small intestine

9

10

8

Note:
Arrows
indicate
direction
blood
flow.

7

6

11

B. **Villus**

One
carpet
fiber

Imagine a folded carpet sample.
The fold itself is like the large fold
or **plica circularis** in the
small intestine while each
individual **carpet fiber** is
like one **villus**.

Folded
carpet sample

Description

The **pancreas** is an elongated, pinkish-colored gland 5–6 inches in length. It is divided into three main sections—**head, body**, and **tail**. The head is the widest portion; the body is the central region; and the tail marks the blunt, tapering end of the gland. The surface is bumpy and nodular. The individual nodes are called **lobules**. Scattered within the gland are many glandular epithelial cells. Also inside is a long tube called the pancreatic duct, which runs through the middle of the gland and terminates in an opening to the duodenum. Smooth muscle surrounds this opening to form a valve called the **hepatopancreatic sphincter**.

The pancreas has a dual function as both an endocrine gland and an exocrine gland. As an exocrine gland, the vast majority of the epithelial cells (99%) form clusters called **acini**. The cells in the acini produce watery secretions and many different digestive enzymes to aid in the process of chemical digestion. These are released into the pancreatic duct and then into the duodenum. As an endocrine gland, the remaining 1% of its cells are arranged in separate cell clusters called pancreatic islets (*islets of Langerhans*). These various cells produce four different hormones that diffuse directly into the bloodstream, travel to various target organs, and induce responses in those organs in order to regulate a variety of different processes in the body.

Analogy

The general **gross anatomy of the pancreas** is like a **tadpole** with a head, body, and tail in one elongated shape. The **surface of the pancreas** has a nodular appearance like that of **alligator skin**. However, its texture is not as coarse.

Location

Abdominal cavity; the head of the pancreas is found near the duodenum and its tail is below and behind the stomach

Function

- For the **digestive system**—Cells produce a wide variety of enzymes in the following categories: **carbohydrases** (*digest carbohydrates*), **lipases** (*digest lipids*), **nucleases** (*digest nucleic acids such as DNA*), and **proteases** (*digest proteins*)

- For the **endocrine system**—Produces the following hormones: **insulin, glucagon, somatostatin**, and **pancreatic polypeptide (PP)**

Key to Illustration

1. Common bile duct
2. Duodenal papilla
3. Duodenum
4. Droplet of either bile or pancreatic juice
5. Hepatopancreatic sphincter
6. Smooth muscle in wall of duodenum
7. Pancreatic duct
8. Body of pancreas
9. Lobules

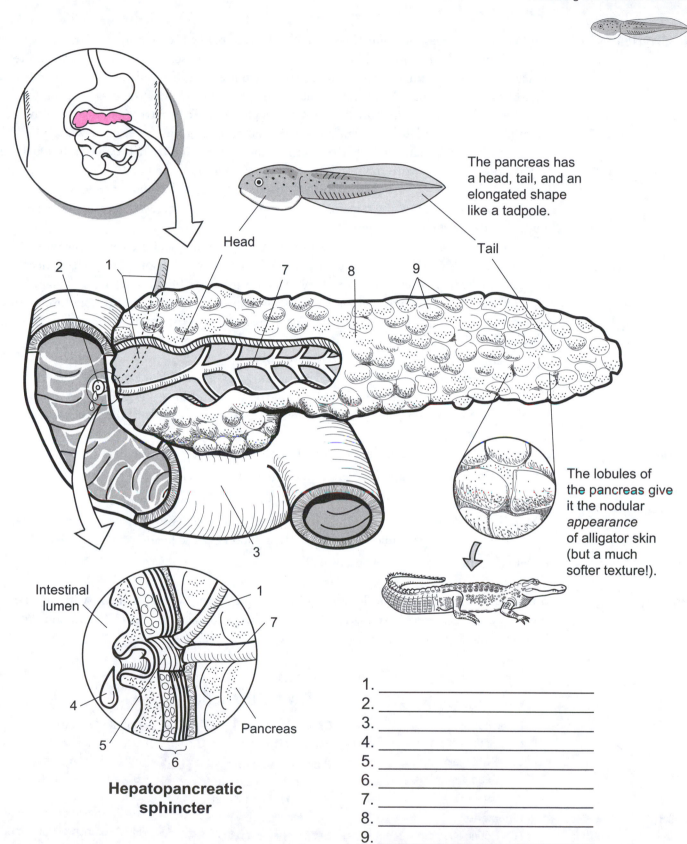

The pancreas has a head, tail, and an elongated shape like a tadpole.

Head

Tail

The lobules of the pancreas give it the nodular *appearance* of alligator skin (but a much softer texture!).

Intestinal lumen

Pancreas

Hepatopancreatic sphincter

1. _____
2. _____
3. _____
4. _____
5. _____
6. _____
7. _____
8. _____
9. _____

Description

The **liver** is the largest abdominal organ and is located below the diaphragm in the abdominal cavity. It is divided into two major lobes—left and right—that are separated by a sheet of connective tissue called the **falciform ligament**. Along the bottom edge of the falciform ligament is a fibrous, cable-like structure called the **round ligament** (*ligamentum teres*), which is derived from the fetal umbilical vein. The liver is anchored to the diaphragm by the **coronary ligament**. The liver also has two minor lobes—**caudate** and **quadrate**—which are visible in the posterior view.

Microscopically, each lobe of the liver is composed of more than 100,000 individual hexagon-shaped units called **hepatic lobules**. These structures are the fundamental functional units of the liver. At each corner of a lobule is a cluster of three vessels called a **hepatic triad**, which consists of a hepatic artery, a branch of the hepatic portal vein, and a bile duct. At the center of each lobule is a long vessel called the **central vein**.

Radiating out from the central vein like spokes from a wheel are the permeable, blood-filled capillaries called **sinusoids**. Fixed to the inner lining of the sinusoids are phagocytic cells called **Kupffer's** cells, which remove most of the bacteria from the blood and digest damaged red blood cells. The hepatic artery and hepatic portal vein deliver blood into the sinusoids and it travels toward the central vein, then into the hepatic veins, and into the inferior vena cava. Most **hepatocytes** (*liver cells*) are located close to the blood in the sinusoids, which makes it easier to deposit products into the blood or screen materials out of it.

Bile is produced by hepatocytes and drains into vessels called **bile canaliculi**, then into the larger **bile ducts**, which transport the bile away from the liver and into the gallbladder for storage. Note that the direction of bile flow in the bile canaliculi is opposite to that of the blood in the sinusoids.

Functions

The liver has more than 200 different functions. Major functions of the liver include:

- **Synthesis**   *ex:* plasma proteins, clotting factors, bile, cholesterol
- **Storage**   *ex:* iron (Fe), glycogen, blood, fat-soluble vitamins
- **Metabolic**   *ex:* convert glucose to glycogen and glycogen to glucose; convert carbohydrates to lipids, maintain normal blood glucose levels
- **Detoxification** *ex:* alcohol and other drugs

Analogy

In cross-section, each **hepatic lobule** is like a simple **Ferris wheel** with six swinging chairs. Each **swinging chair** is a **hepatic triad**. The **hub** of the Ferris wheel is the **central vein**, and the **spokes** of the Ferris wheel are like the **sinusoids**.

Key to Illustration

Liver Lobes (L)
L1. Left major
L2. Right major
L3. Caudate
L4. Quadrate

Ligaments
RL. Round ligament
    (*Ligamentum teres*)
FL. Falciform ligament
CL. Coronary ligament

Other
GB. Gallbladder

Blood Vessels (B)
B1. Inferior vena cava
B2. Hepatic portal vein
B3. Hepatic artery proper
B4. Left hepatic vein
B5. Opening to right hepatic vein

Lobule Structures
CV. Central vein
HT. Hepatic triad
BC. Bile canaliculus
BD. Bile duct
HA. Branch of hepatic artery
HPV. Branch of hepatic portal vein
S. Sinusoid
H. Hepatocyte

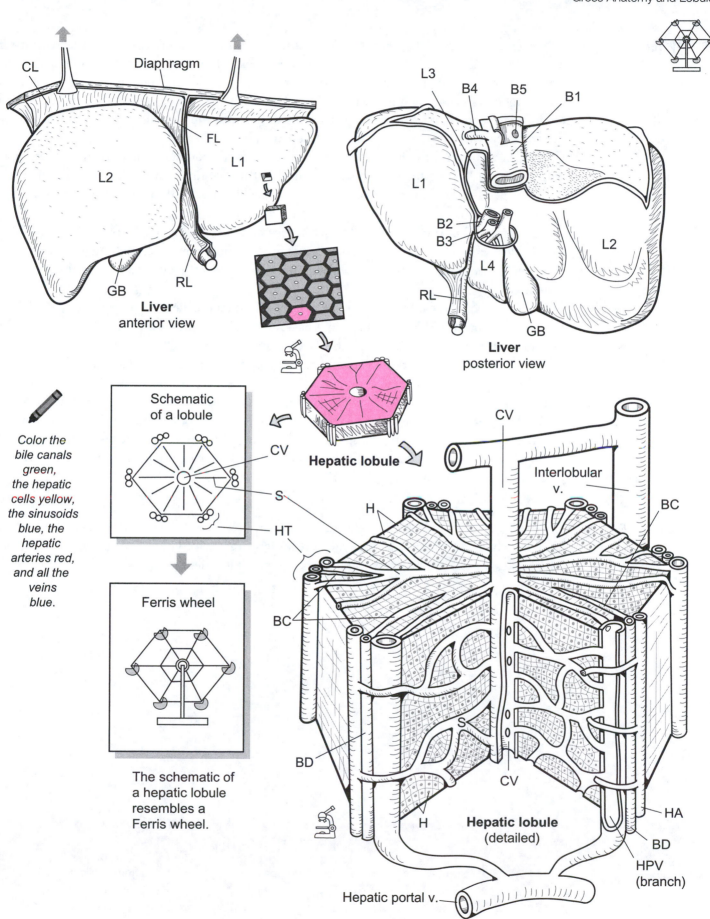

CL
Diaphragm
FL
L2
L1
RL
GB

Liver
anterior view

L3
B4
B5
B1
L1
B2
B3
L4
RL
L2
GB

Liver
posterior view

Color the bile canals green, the hepatic cells yellow, the sinusoids blue, the hepatic arteries red, and all the veins blue.

Schematic of a lobule

CV
S
HT

Hepatic lobule

Ferris wheel

The schematic of a hepatic lobule resembles a Ferris wheel.

CV
Interlobular v.
BC
H
BC
S
BD
CV
H
Hepatic lobule
(detailed)
HA
BD
HPV
(branch)

Hepatic portal v.

Description

Bile is produced by the **hepatocytes** (*liver cells*), stored in the **gallbladder**, and released into the **duodenum**. It is composed mostly of water, some ions, bilirubin (protein pigment), and various lipids. After being produced, it drains away from the hepatic lobule and enters a microscopic **bile duct**, which drains into a larger **hepatic duct**. The **left hepatic duct** drains bile from the left lobe of the liver, and the **right hepatic duct** drains bile from the right lobe of the liver. The two hepatic ducts then fuse to form a **common hepatic duct**.

Bile flows down the common hepatic duct and into the **common bile duct**, which terminates in an opening leading to the duodenum. Around this opening is a thickened band of smooth muscle called the **hepatopancreatic sphincter**. Normally the smooth muscle around this sphincter is contracted, causing it to be closed.

Bile is stored in the **gallbladder**, which is located under the right lobe of the liver. To fill the gallbladder with bile, the hepatopancreatic sphincter must remain closed. As bile drains from the liver, it backs up into the common bile duct and enters the cystic duct, which leads to the gallbladder.

Emptying of the gallbladder is regulated by a hormone called **cholecystokinin (CCK)**. As lipid-rich **chyme** (food and gastric juice) passes through the **mucosa** (inner lining) of the duodenum, it stimulates specific cells in the duodenum to release CCK. This hormone then travels through the bloodstream and targets the smooth muscle around the gallbladder.

These smooth muscle cells have receptors for CCK. Once CCK binds to these receptors, it induces the smooth muscle to contract. The result is that bile is expelled forcefully from the gallbladder. Simultaneously, CCK targets the smooth muscle of the hepatopancreatic sphincter and causes it to relax, opening the sphincter. With the sphincter open, bile moves out of the gallbladder, down the cystic duct, down the common bile duct, past the hepatopancreatic sphincter, and into the duodenum.

Function

Emulsification of lipids (fats) is the first step in the chemical digestion of lipids. Rather than breaking chemical bonds, bile physically breaks large masses of lipids into smaller lipid droplets. This makes it easier for enzymes such as **lipase** to facilitate chemical digestion.

Key to Illustration

1. Right hepatic duct
2. Left hepatic duct
3. Common hepatic duct
4. Cystic duct
5. Common bile duct
6. Pancreatic duct

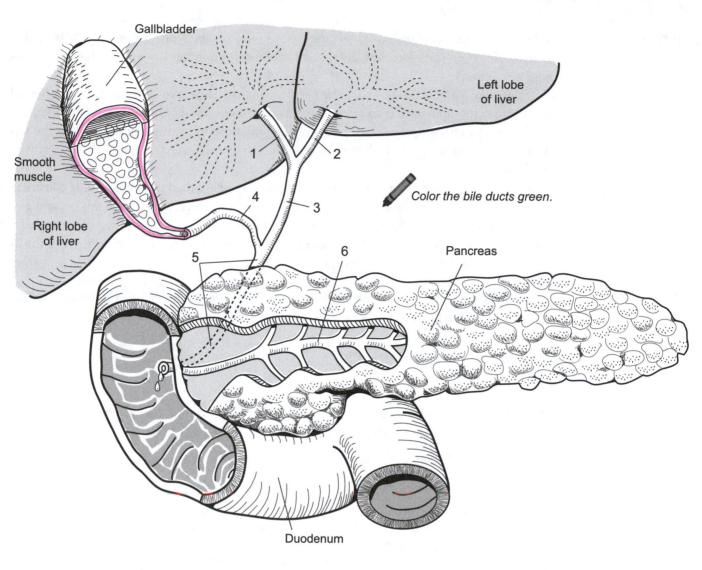

Gallbladder

Left lobe of liver

Smooth muscle

Right lobe of liver

1

2

4

3

5

6

Pancreas

Color the bile ducts green.

Duodenum

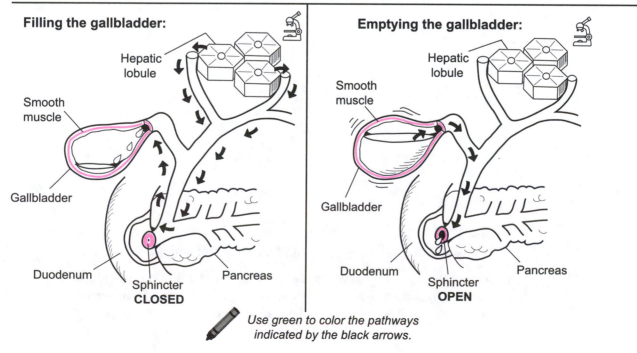

Filling the gallbladder:

Hepatic lobule

Smooth muscle

Gallbladder

Duodenum

Sphincter **CLOSED**

Pancreas

Emptying the gallbladder:

Hepatic lobule

Smooth muscle

Gallbladder

Duodenum

Sphincter **OPEN**

Pancreas

Use green to color the pathways indicated by the black arrows.

Description

The **large intestine**, or large bowel, is a hollow, muscular tube about 5 feet long. It is subdivided into three different parts—**cecum**, **colon**, and **rectum**. It connects the end of the small intestine (*ileum*) to the **anus**.

The **cecum** is a pouch that marks the beginning of the large intestine. On the posterior side of this structure, the **appendix** (*vermiform appendix*) can be located. This is a long, slender, hollow tube-like structure that opens into the cecum. The next portion is the **colon**, which is the longest part of the large intestine. It is subdivided into four segments—*ascending*, *transverse*, *descending*, and *sigmoid* colon.

Along the length of the colon is a band of smooth muscle called the **tenia coli**. It constricts the colon into pouches called **haustra** that run all along its length and permit expansion and elongation of the colon. At regular intervals along the taenia coli are flaps of fatty tissue called **fatty appendices** (*epiploic appendages*). The **rectum** is the last segment of the large intestine and the digestive tract and allows for temporary storage of fecal waste.

Analogy

The **appendix** is also called the *vermiform appendix*. The term *vermiform* means "worm-like" so the appendix is compared to a **worm** because of its general shape.

Location

Abdominal cavity; surrounds the small intestine

Functions

- **Reabsorption** of water and electrolytes

- **Absorption** of some vitamins (*e.g.*, vitamin K, B-complex vitamins) produced by the bacteria *Escherichia coli* (E. coli), which naturally live within the colon.

- **Compaction and temporary storage** of fecal waste

Key to Illustration

1. Cecum	4. Left colic (*splenic*) flexure	7. Ileocecal valve
2. Right colic (*hepatic*) flexure	5. Tenia coli	8. Rectum
3. Haustra	6. Fatty appendages	9. Anus

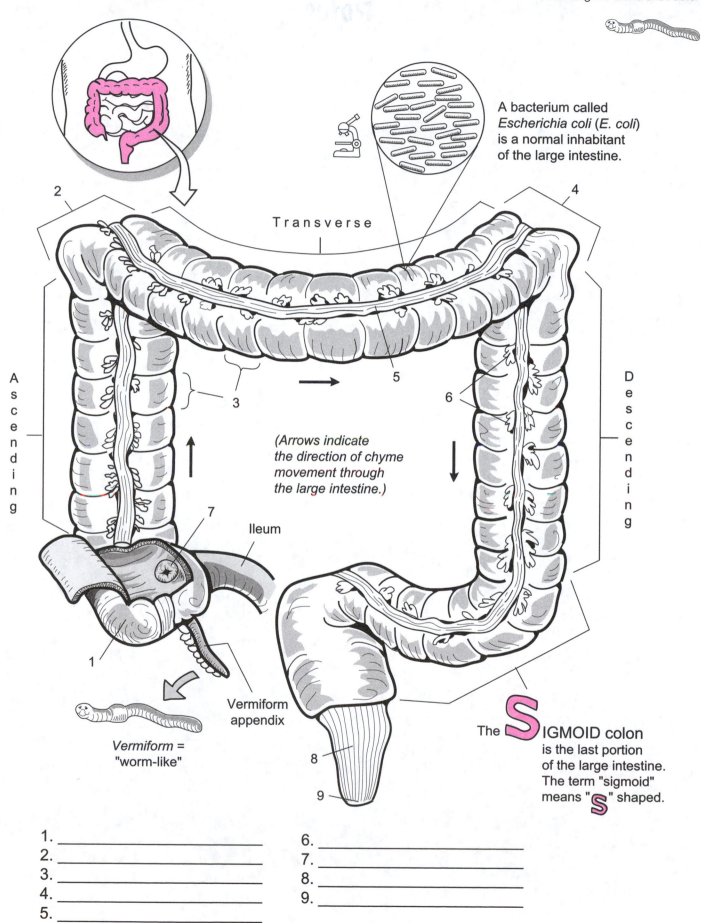

A bacterium called
Escherichia coli (*E. coli*)
is a normal inhabitant
of the large intestine.

Transverse

A
s
c
e
n
d
i
n
g

D
e
s
c
e
n
d
i
n
g

*(Arrows indicate
the direction of chyme
movement through
the large intestine.)*

Ileum

Vermiform
appendix

*Vermiform =
"worm-like"*

The **S**IGMOID colon
is the last portion
of the large intestine.
The term "sigmoid"
means "**S**" shaped.

1. _____
2. _____
3. _____
4. _____
5. _____

6. _____
7. _____
8. _____
9. _____

Notes

Urinary System

Description

The urinary system is composed of the **kidneys**, **ureters**, **urinary bladder**, and **urethra**. The kidneys are located on either side of the upper lumbar region of the vertebral column. They lie between the dorsal body wall and the **parietal peritoneum** in a retroperitoneal position. They are tightly covered by the parietal peritoneum on their anterior surface and further protected by layers of fatty tissue. Major abdominal organs such as the small and large intestines are all positioned anterior to the kidneys.

The **renal arteries** branch off the **abdominal aorta** and supply oxygenated blood to the kidneys, while the **renal veins** drain blood from the kidneys and empty into the **inferior vena cava**. The blood that enters the kidney via the renal artery contains various waste products that must be removed from the blood. The functional units within the kidney are microscopic structures called **nephrons**. They filter waste products and other substances from the blood and transfer them to a separate tubular system. The fluid that enters the tubular system is called the filtrate. The filtrate is processed in a manner that transports nutrients back to the bloodstream and retains waste products in the tubular system.

When the processing is completed, the liquid inside the tube is called urine. In reality, urine is nothing more than processed blood plasma. The urine then moves out of the nephrons and into the pelvis of the kidneys. Slender, muscular tubes called **ureters** transport urine from the kidney to the urinary bladder. The muscular **urinary bladder** can expand to hold as much as 600–800 ml. of urine. Finally, urine is removed from the body by the process of **micturition** (*urination*), which uses forceful muscular contractions in the wall of the urinary bladder to transport urine into the urethra and out of the body.

Key to Illustration

Major Organs/Structures

1. Kidney
2. Ureter
3. Urinary bladder

Blood Vessels (B)

B1. Inferior vena cava
B2. Abdominal aorta
B3. Renal a.
B4. Renal v.
B5. Common iliac a.

B6. Internal iliac a.
B7. External iliac a.
B8. Gonadal v. *(testicular v. in male; ovarian v. in female)*
B9. Gonadal a. *(testicular a. in male; ovarian a. in female)*

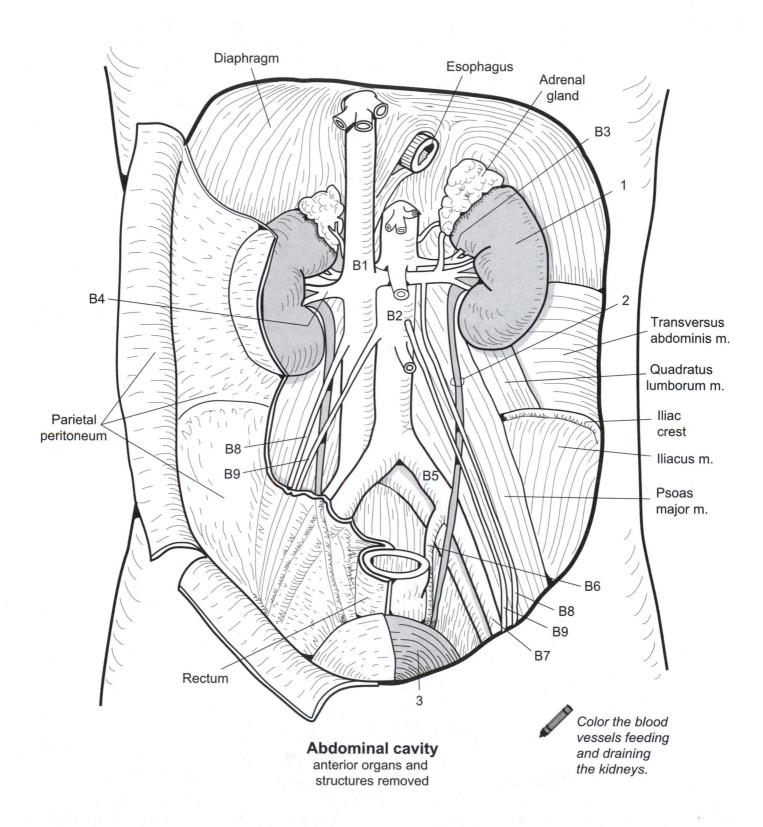

Diaphragm

Esophagus

Adrenal gland

B3

1

B1

B2

B4

Transversus abdominis m.

Quadratus lumborum m.

Iliac crest

Iliacus m.

Psoas major m.

2

Parietal peritoneum

B8

B9

B5

B6

B8

B9

B7

Rectum

3

Abdominal cavity
anterior organs and
structures removed

*Color the blood
vessels feeding
and draining
the kidneys.*

Description

The kidneys are covered by a fibrous tissue called the **renal capsule**. The **renal hilus**, a cleft or indentation on the medial surface of the kidneys, is a handy landmark for locating the **ureters**, blood vessels (*renal a., renal v.*), and nerves that serve this organ. Each kidney is divided into three regional areas: cortex, medulla, and pelvis. The **cortex** is the outermost region, like the bark around a tree, and appears granular. The darker colored **medulla** is the middle region, containing numerous funnel-shaped structural units called **renal pyramids,** each with a **papilla** at its tapered end.

Separating the pyramids are projections of cortical tissue called **renal columns**. The central region is the **pelvis**—a pouch that narrows and extends directly into the **ureter**.

Near the medulla, the pelvis branches off into structures called **calyces** (sing., *calyx*). Each cup-shaped **minor calyx** surrounds each papillus of a renal pyramid to receive urine from it. When several of these minor calyces join together, they form a larger chamber called a **major calyx**. As urine is collected in the pelvis, it moves down the **ureter** and into the urinary bladder.

Analogy

The **calyces** (sing., *calyx*) are functionally like a **plumbing system** of smaller-diameter pipes leading to larger ones. Each **smaller pipe** is a **minor calyx**, and each **larger pipe** is a **major calyx**. The flow of water through the pipes follows a path similar to the flow of urine through the calyces.

Key to Illustration

Regional Areas
1. Cortex
2. Medulla
3. Pelvis

Other Structures
4. Renal papillus
5. Renal pyramids
6. Renal capsule
7. Renal column
8. Minor calyx
9. Major calyx
10. Adipose tissue in renal sinus
11. Renal hilus
12. Ureter

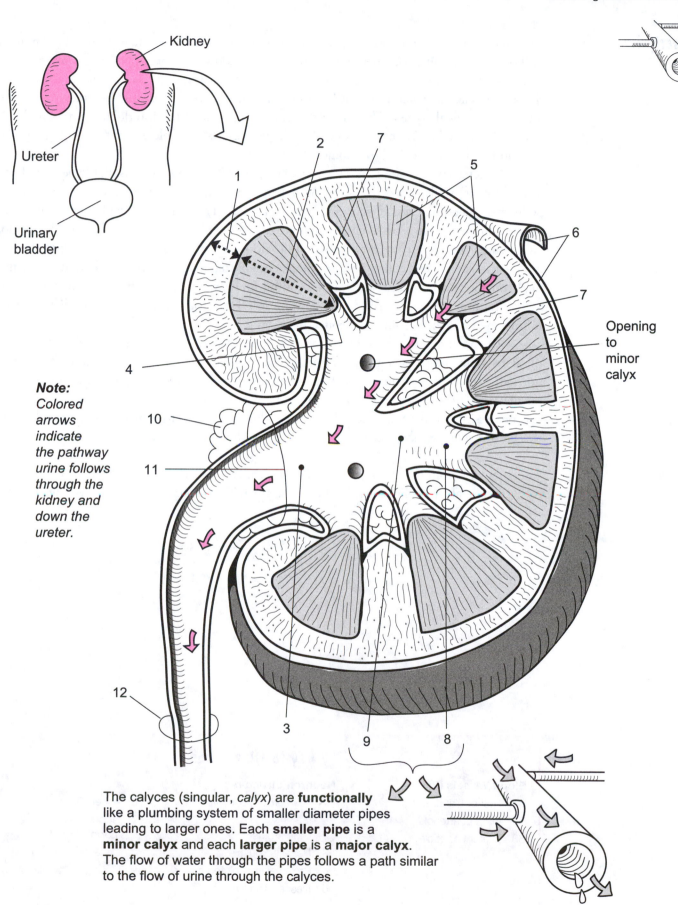

Kidney

Ureter

Urinary
bladder

1

2

7

5

6

7

Opening
to
minor
calyx

4

Note:
*Colored
arrows
indicate
the pathway
urine follows
through the
kidney and
down the
ureter.*

10

11

12

3

9

8

The calyces (singular, *calyx*) are **functionally**
like a plumbing system of smaller diameter pipes
leading to larger ones. Each **smaller pipe** is a
minor calyx and each **larger pipe** is a **major calyx**.
The flow of water through the pipes follows a path similar
to the flow of urine through the calyces.

Description

The **nephron** is the microscopic, functional unit of the kidney. Each kidney has more than a million of these units. Each nephron is divided into six distinct parts: *glomerulus*, *glomerular capsule*, *proximal convoluted tubule*, *nephron loop*, *distal convoluted tubule*, and *collecting tubule*. As blood enters the kidney via the **renal artery**, it branches into smaller vessels until it becomes the interlobular artery in the **renal cortex**. The afferent arteriole branches off the interlobular artery and connects to a special coiled ball of capillaries called the **glomerulus**. Completely surrounding the glomerulus is a cup-like structure called the **glomerular capsule**.

Connected to the capsule is a long tubular system. The glomerulus is highly permeable because of pores in its walls called **fenestrae**. Wrapped around the glomerular capillaries are cells called **podocytes**—modified simple squamous epithelial cells. The blood in the glomerulus is under high pressure so the plasma is passively filtered into the glomerular capsule. The fluid inside the glomerular capsule is referred to as *filtrate*. This filtrate flows from the glomerular capsule into the first part of the tubular system called the **proximal convoluted** (*coiled*) **tubule**. Next it flows into the **nephron loop**, which consists of a descending limb, where the filtrate flows downward and an ascending limb where the filtrate moves upward. Then the filtrate enters the **distal convoluted tubule** and finally moves into the **collecting tubule**.

The general function of the nephron is to collect the filtrate (*filtered blood plasma*) in a separate tubular system and process it. The filtrate in the glomerular capsule is a solution that contains both nutrients and waste products mixed together. The nephron processes the filtrate by returning the nutrients to the bloodstream and retaining the waste products in the tubular system. By the time this processing job is completed, the resulting fluid is referred to as **urine**. The urine then follows this pathway: *minor calyx*, *major calyx*, *ureter*, *urinary bladder*, *urethra*, and it exits the body.

Analogy

The tubular system in the nephron is **structurally** like the sewer lines in private homes that connect to a larger public sewer drain under a street. The **sewer drain** is like the **collecting tubule**.

Key to Illustration

Blood Vessels (B)

B1. Interlobular artery
B2. Afferent arteriole
B3. Efferent arteriole

Nephron Structures

1. Glomerular (*Bowman's*) capsule
2. Glomerulus
3. Renal corpuscle
4. Proximal convoluted tubule
5. Descending limb
6. Ascending limb
7. Nephron loop (*Loop of Henle*)
8. Distal convoluted tubule
9. Collecting tubule (*duct*)

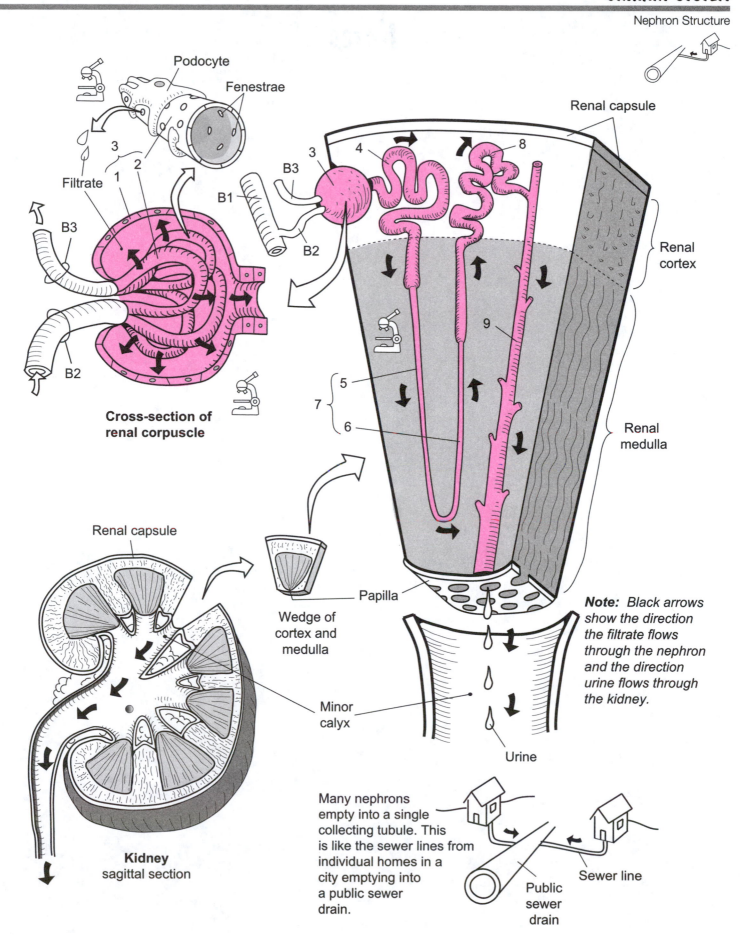

Podocyte

Fenestrae

3

2

1

Filtrate

B3

B2

Cross-section of renal corpuscle

B3

B1

3

B2

4

8

Renal capsule

Renal cortex

9

5

7

6

Renal medulla

Renal capsule

Wedge of cortex and medulla

Papilla

Minor calyx

Kidney sagittal section

Note: *Black arrows show the direction the filtrate flows through the nephron and the direction urine flows through the kidney.*

Urine

Many nephrons empty into a single collecting tubule. This is like the sewer lines from individual homes in a city emptying into a public sewer drain.

Public sewer drain

Sewer line

Notes

Reproductive Systems

Detailed Pathway   **Simplified Pathway**

1. Lumen of seminiferous tubule   **S**eminiferous tubule

2. Rete testis   **R**ete Testis

3. Efferent ductules   **E**fferent ductules

4. Epididymis   **E**pididymis

5. Ductus (*vas*) deferens   **D**uctus deferens

6. Ampulla of ductus deferens

7. Ejaculatory duct   **E**jaculatory duct

8. Prostatic urethra

9. Membranous urethra   **U**rethra

10. Penile urethra

11. External urethral orifice

Study Tip

Use this mnemonic for the simplified pathway shown above:

"**S**ome **R**eally **E**lderly **E**lephants **D**on't **E**ven **U**rinate!"

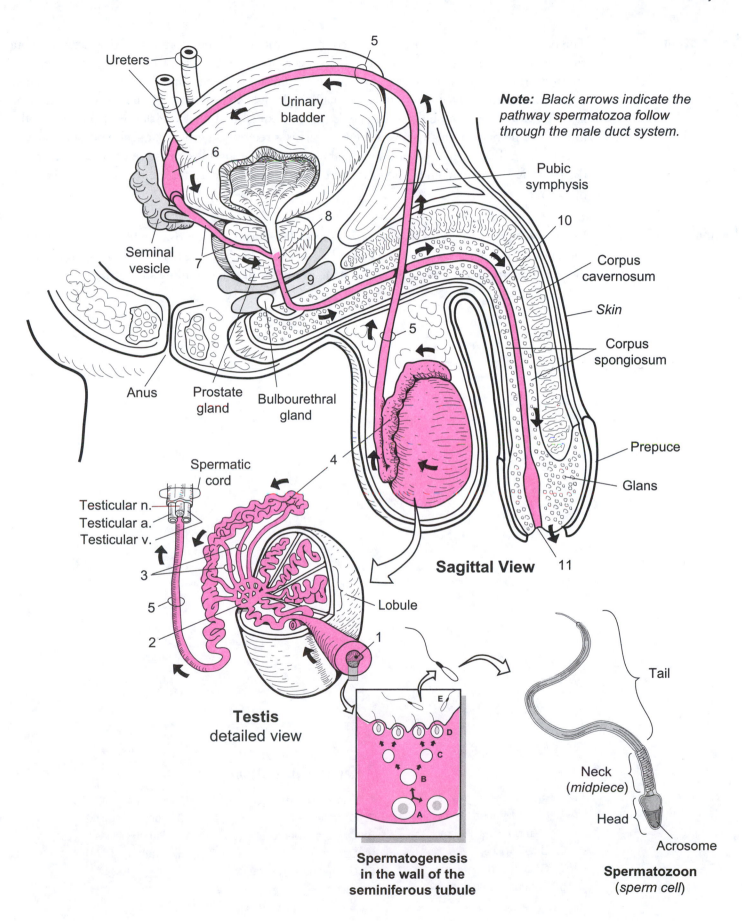

Ureters

Urinary
bladder

5

6

8

7

9

Seminal
vesicle

Anus

Prostate
gland

Bulbourethral
gland

Note: Black arrows indicate the
pathway spermatozoa follow
through the male duct system.

Pubic
symphysis

10

Corpus
cavernosum

Skin

Corpus
spongiosum

Prepuce

Glans

5

Sagittal View

11

4

Lobule

Spermatic
cord

Testicular n.

Testicular a.

Testicular v.

3

5

2

1

Testis
detailed view

E

D

C

B

A

**Spermatogenesis
in the wall of the
seminiferous tubule**

Tail

Neck
(*midpiece*)

Head

Acrosome

Spermatozoon
(*sperm cell*)

Description

The penis, the male sex organ, is composed of three tubes of spongy connective tissue: two corpora cavernosa and one corpus spongiosum. The two **corpora cavernosa** constitute the bulk of the penis and rest on top of the **corpus spongiosum**, which surrounds the urethra. Sexual stimulation causes these tubes to fill with blood during a normal erection. The corpus spongiosum becomes the **glans** at its distal end and the **bulb** of the penis at its proximal end. A loose sleeve of skin called the foreskin or prepuce normally covers the glans. This is removed during a surgical procedure called a circumcision. Each corpus cavernosum (*plural, corpora cavernosa*) becomes the **crus** (*plural crura*) of the penis at its proximal end. These crura are part of an attachment for the root of the penis to the pubic arch in the pelvis.

Analogy

In cross-section, the penis is like a monkey's face. The **corpus cavernosa** are like the **mask around the eyes** of the monkey's face. The **central arteries** are like the **eyes**. The **corpus spongiosum** is like the **area around the mouth** and the **male urethra** is like the **monkey's mouth**.

Key to Illustration

1. Bulbourethral *(Cowper's glands)* glands
2. Membranous urethra
3. Bulb of penis
4. Crus of penis
5. Subcutaneous dorsal v.
6. Deep dorsal a.
7. Deep fascia
8. Corpora cavernosa
9. Central a.
10. Corpus spongiosum
11. Glans
12. External urethral orifice
13. Tunica albuginea
14. Superficial fascia
15. Urethra

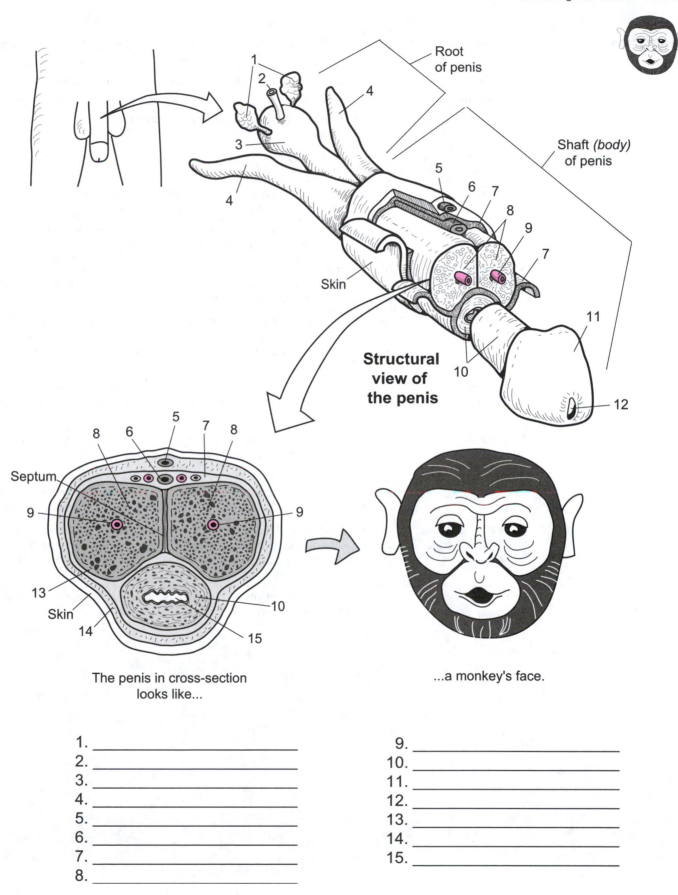

Root of penis

Shaft (body) of penis

Skin

Structural view of the penis

Septum

Skin

The penis in cross-section looks like...

...a monkey's face.

1. _____
2. _____
3. _____
4. _____
5. _____
6. _____
7. _____
8. _____

9. _____
10. _____
11. _____
12. _____
13. _____
14. _____
15. _____

Description

The female **external genitalia** (*vulva*) are illustrated on the facing page. Please understand that this is an idealized rendering and does not account for the many variations of normal. In the developing embryo, the external genitalia of males and females appear very similar. Then they differentiate over a period of about 8 weeks. The table will point out *homologous structures*—features that are structurally similar between male and female genitalia.

Structure	Description
1. Mons pubis	A relatively large mound of skin and fatty tissue located anterior to the pubic symphysis. It's covered with pubic hair in adult females
2. Prepuce of clitoris	A hood-like structural extension of the labia minora that covers the glans of the clitoris. This is homologous to the male prepuce (*foreskin*).
3. Glans of clitoris	The tip of the clitoris that contains many sensory nerve endings for sexual pleasure in the female. This is homologous to the glans of the penis. The clitoris is a small, erectile body that engorges with blood during sexual excitation.
4. Urethral orifice (*opening*)	Opening from the urethra located between the glands of the clitoris and the vaginal opening. The urethra is a narrow tube that connects the urinary bladder to outside of the body. Urine collects in the urinary bladder, passes through the urethra, and is expelled from the body
5. Labia minora ("smaller lips")	Smaller, hairless folds located inside the larger labia majora that may have increased pigmentation due to the abundance of melanocytes. They are homologous to the ventral shaft of the penis.
6. Vestibule	The space between the labia minora that contains the urethral opening, the vaginal orifice, and openings to the greater vestibular glands.
7. Vaginal orifice (*opening*)	The opening into the vagina. The vagina is a thick, muscular tube that connects the uterus to outside the body. It acts as the organ to receive the penis during sexual intercourse. It also functions as the birth canal and passageway for menstruation.
8. Openings for the greater vestibular glands	The openings that lead to the pair of greater vestibular glands. During sexual arousal, these glands produce a secretion that serves as a vaginal lubricant. Secretion increases during sexual intercourse. These glands are homologous to the bulbourethral glands in males.
9. Labia majora ("larger lips")	Thick, protruding folds of fatty skin that are homologous to the male scrotum. Outer margins are covered with coarse pubic hair in the adult female.

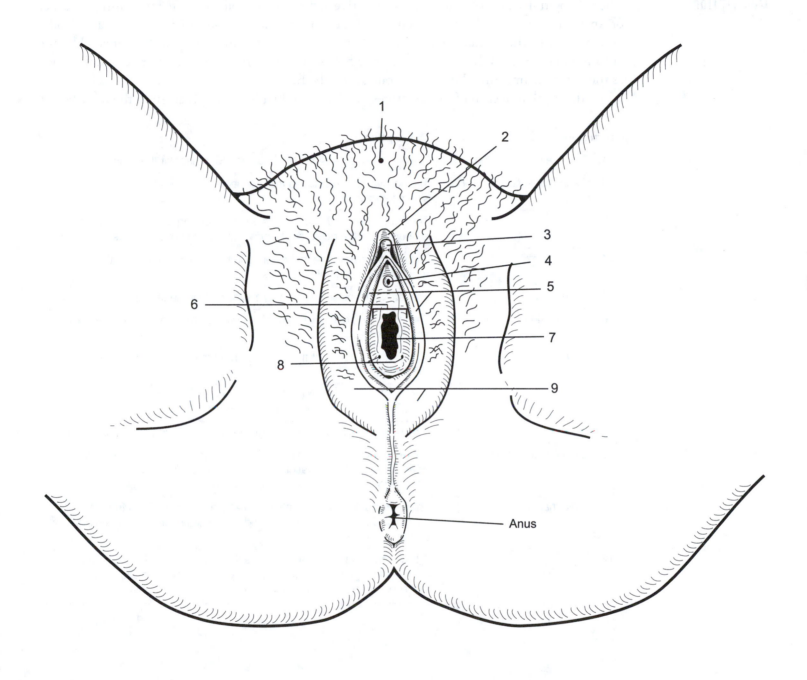

Anus

Color the major structures different colors.

1. _____

2. _____

3. _____

4. _____

5. _____

6. _____

7. _____

8. _____

9. _____

Description

The midsagittal view of the female reproductive system allows you to see its relationship to other organs/structures in the body. For example, notice that the uterus is superior to the urinary bladder and anterior to the rectum. Moreover, it allows you to see structures in their full length like the clitoris and the vagina. It also helps illustrate how structures connect one structure to another such as the vagina connecting the uterus to outside the body.

A description of each of the structures numbered in the illustration is given in the table below.

Structure	Description
1. Fimbriae of the uterine tube	The finger-like extensions of the uterine tube nearest the ovary
2. Uterine tube (*fallopian tube, oviduct*)	The hollow, muscular tubes that connect the ovaries to the uterus; the site of fertilization
3. Ovary	The female gonad that produces an ovum or egg cell; two ovaries are suspended in the pelvic cavity, one on each side of the uterus
4. Perimetrium	The serous membrane that extends from the peritoneal lining that covers most of the outside of the uterus
5. Myometrium	The thick, muscular layer that forms the wall of the uterus
6. Endometrium	The innermost, glandular layer of the uterus; site where the developing embryo implants
7. Cervix	A neck-like structure at the inferior portion of the uterus that projects into the vagina
8. Mons pubis	A relatively large mound of skin and fatty tissue located anterior to the pubic symphysis; covered with pubic hair in adult females.
9. Clitoris	A small, erectile body that engorges with blood during sexual excitation.
10. Labia minora ("smaller lips")	Smaller, hairless folds located inside the larger labia majora that may have increased pigmentation due to the abundance of melanocytes; homologous to the ventral shaft of the penis.
11. Labia majora ("larger lips")	Thick, protruding folds of fatty skin that are homologous to the male scrotum. Outer margins are covered with coarse pubic hair in the adult female.
12. Vagina	A thick muscular tube that connects the uterus to outside the body; acts as the organ to receive the penis during sexual intercourse and also functions as the birth canal and the passageway for menstruation.

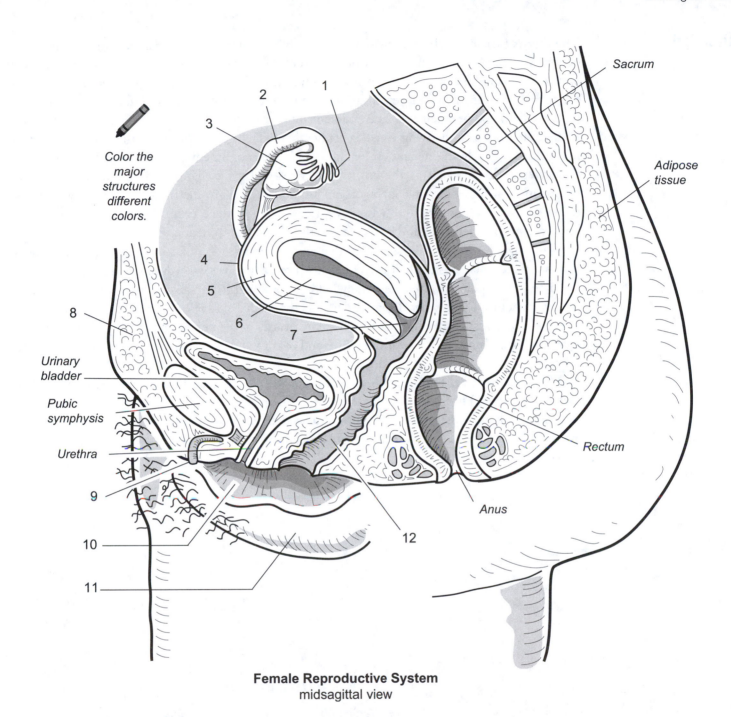

Color the major structures different colors.

Sacrum

Adipose tissue

Urinary bladder

Pubic symphysis

Urethra

Rectum

Anus

Female Reproductive System
midsagittal view

1. _____
2. _____
3. _____
4. _____
5. _____
6. _____

7. _____
8. _____
9. _____
10. _____
11. _____
12. _____

Description

The **uterus** (*womb*) is a hollow, muscular organ divided into three regional areas: fundus, body, and cervix. The **fundus** is the most superior portion of the uterus, the main portion is called the **body**, and the narrowed, neck-like portion that extends into the vagina is called the **cervix**. The wall of the uterus is made of three layers, from innermost to outermost, and are as follows: endometrium, myometrium, and perimetrium. The **endometrium**, or mucosal lining, is made of simple columnar epithelium and an underlying vascular connective tissue. This entire layer thickens during a normal menstrual cycle in preparation for the implantation of an embryo. If no implantation occurs, this layer is sloughed off at the end of the menstrual cycle. The thick middle layer, the **myometrium**, is made of multiple layers of smooth muscle. The myometrium is hormonally stimulated to contract during childbirth to help move the baby out of the uterus. The outermost layer, the fibrous connective **perimetrium,** is the same as the visceral peritoneum.

The **vagina** is a thick, muscular tube that extends from the cervix to outside the body. It is lined with stratified squamous epithelium. The penis enters this passageway during sexual intercourse.

The small, lumpy **ovaries** are loosely held in place within the abdomen by various connective tissue ligaments. The **ovarian ligament** anchors the medial side of each ovary to the lateral side of the uterus, the **suspensory ligament** anchors the ovaries to the pelvic wall, and the wide, flat, **broad ligament** extends like a tarp over the uterus and ovaries and cradles the vagina, uterus, and uterine tubes.

The **uterine tubes** (consisting of the *Fallopian tubes* and *oviducts*) serve to transport a female gamete (*egg cell; ova*) from the ovary to the uterus. The proximal end (near the uterus) is a long, narrow tube called the **isthmus**. The distal end widens and curves around the ovary to form the **ampulla**. The ampulla becomes the funnel-shaped **infundibulum,** which has fingerlike extensions called **fimbriae**.

Key to Illustration

Uterus
1. Fundus of uterus
2. Body of uterus
3. Cervix
4. Lumen of uterus
5. Endometrium
6. Myometrium
7. Perimetrium
8. Internal os
9. Cervical canal
10. External os

Vagina (V)

Blood Vessels (B)
B1. Ovarian artery
B2. Ovarian vein

Ligaments (L)
L1. Round ligament
L2. Suspensory ligament
L3. Ovarian ligament
L4. Broad ligament

Ovary (O)

Uterine Tubes (U)
U1. Ampulla
U2. Isthmus
U3. Infundibulum
U4. Fimbriae

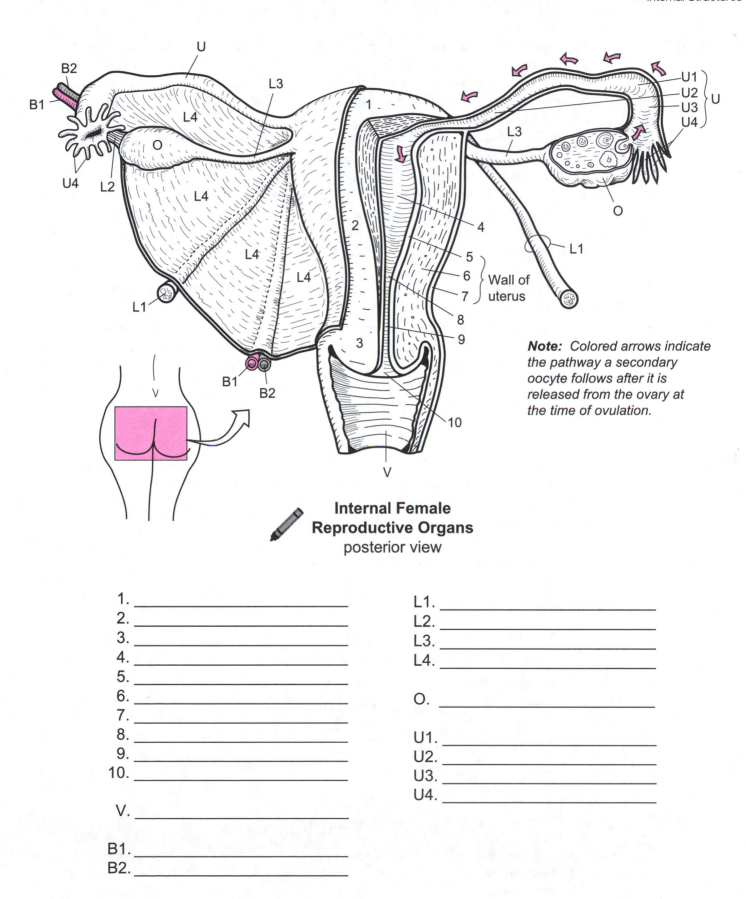

Note: Colored arrows indicate the pathway a secondary oocyte follows after it is released from the ovary at the time of ovulation.

Internal Female Reproductive Organs
posterior view

1. _____
2. _____
3. _____
4. _____
5. _____
6. _____
7. _____
8. _____
9. _____
10. _____

V. _____

B1. _____
B2. _____

L1. _____
L2. _____
L3. _____
L4. _____

O. _____

U1. _____
U2. _____
U3. _____
U4. _____

Description

The female has two lumpy, oval-shaped structures called **ovaries**. Each measures about 5 cm. long. Blood is brought to the ovary by an **ovarian artery** and drained by an **ovarian vein**. The surface of the ovary is covered by a simple cuboidal epithelial layer called the **germinal epithelium**. The two regional areas are the outer **cortex** and the inner **medulla**. The production of **gametes** (*sex cells*) occurs in the **cortex**.

Ideally, the ovaries in a sexually mature female alternate to produce one ovum each month. The specific process for developing an **ovum** (*egg cell*) is called **oogenesis**. This is a part of the ovarian cycle, which refers to all the processes that occur in the ovary during this monthly event. The ova are produced in special chambers called **follicles**. The process begins when a hormone stimulates an immature follicle, **primordial follicle**, to mature. The wall of the follicle thickens and cells within it begin to produce a fluid called **follicular fluid,** which fills a space called the **antrum**. As the follicle expands, it goes through the following progression:

Primordial follicle ⟶ Primary follicle ⟶ Secondary follicle ⟶ Tertiary follicle

Once a **tertiary** (*Graafian, Vesicular*) **follicle** has been formed, a hormonal surge causes it to rupture and an ovum is released in an event called **ovulation**. The specific name for this ovum is a **secondary oocyte**. The broken tissue within the tertiary follicle thickens and becomes a temporary endocrine gland called a **corpus luteum,** which produces a mixture of **estrogens** and **progestins**. Gradually, the corpus luteum shrinks and degrades into a small mass of scar tissue called the **corpus albicans**.

Analogy

The **maturation of a follicle** in the ovary is compared to a **water balloon filling with water**. As follicular fluid accumulates inside the antrum, it increases pressure just like water inside a water balloon. When the pressure becomes too great, the follicle ruptures at the **moment of ovulation** just like the **bursting of a water balloon.**

Location

Ovaries are located along the lateral edges of the pelvic cavity.

Function

The ovaries have two basic functions:

- Produce **gametes** (sex cells)

- Manufacture and release **hormones**

    Follicles (primary, secondary, and tertiary) ⟶ **estrogen**

    Corpus luteum ⟶ **estrogen, progesterone**

Key to Illustration

1. Primary follicle
2. Secondary follicle
3. Tertiary follicle
4. Ovulation
5. Early corpus luteum
   (*still forming*)
6. Corpus luteum
7. Corpus albicans

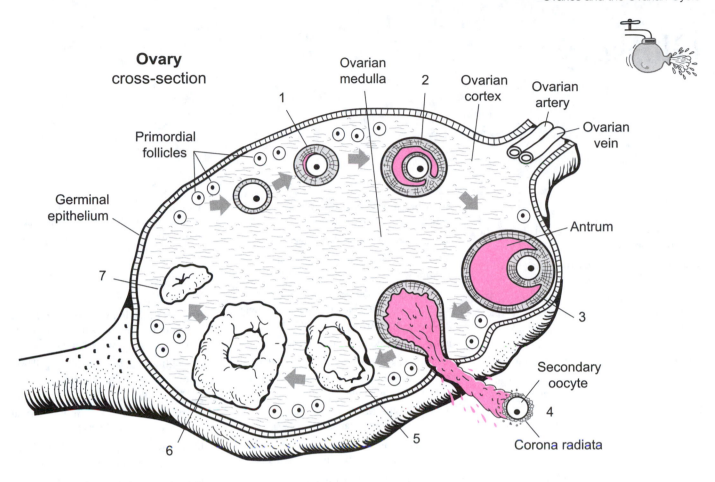

Ovary
cross-section

Ovarian medulla

Ovarian cortex

Ovarian artery

Ovarian vein

Primordial follicles

Germinal epithelium

Antrum

Secondary oocyte

Corona radiata

1 2 3 4 5 6 7

The maturation process for a follicle is like...

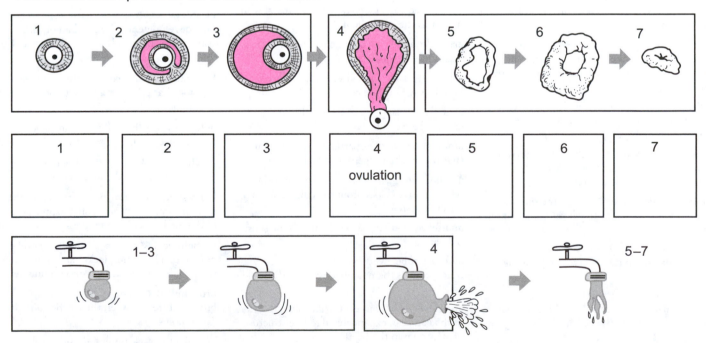

1 2 3 4 5 6 7

1 2 3 4 ovulation 5 6 7

1–3 4 5–7

...a water balloon filling up with water and then bursting.

Glossary

A

abdomen AB-doh-men the region between the diaphragm and the pelvis

absorption ab-ZORP-shun the passage of gases, liquids or solutes through a membrane

acidic ah-SID-ik describes a solution where the pH is below 7; having a relatively high concentration of hydrogen ions

acromial ah-KROH-mee-al the outer end of the scapula; site where the clavicle is attached

acrosome AK-roh-sohm caplike structure at the end of sperm; produces enzymes for egg penetration

actin AK-tin a thin, protein filament found in skeletal muscle cells; protein component of microfilaments

adipocyte AD-i-poh-syte a fat cell

afferent AF-fer-ent toward; opposite of efferent

alkaline AL-kah-lin basic; pH greater than 7; having a relatively low concentration of hydrogen ions

alveolus/alveoli al-VEE-oh-luss/al-VEE-oh-lye a delicate air sac at the end of the bronchial tree in the lungs; where gas exchange occurs with the blood

amino acid ah-MEE-no ASS-id the structural unit of a protein composed of carbon, hydrogen, oxygen, and nitrogen

amnion AM-nee-on one of the extraembryonic membranes; develops around the embryo/fetus forming the amniotic cavity

amniotic fluid am-nee-OT-ik fluid the fluid that surrounds and cushions the developing embryo and fetus

amniotic sac am-nee-OT-ik sak fluid-filled chamber in which the embryo floats during development

anaphase AN-nah-fayz stage of mitosis; when the chromatid pairs separate and move toward the opposite ends of the cell

antebrachial an-tee-BRAY-kee-al pertaining to the forearm

antecubital an-tee-KYOO-bi-tal pertaining to the anterior side of the elbow

anterior an-TEER-ee-or toward the front or ventral; opposite of posterior

antrum AN-trum central chamber

anus AY-nus external opening at the end of the rectum

aorta ay-OR-tah largest artery in the body; carries oxygenated blood from left ventricle and into the systemic circuit

apex AY-peks pointed tip of a structure

aponeurosis ap-oh-nyoo-ROH-sis broad, flat collagenous sheets that may serve as anchor points for skeletal muscle

appendix ah-PEND-diks small organ connected to the cecum of the large intestine

aqueous humor AY-kwee-us HYOO-mor a fluid that fills the anterior chamber of the eye

arachnoid ah-RAK-noyd middle meninges that surround the CSF and protect the brain and spinal cord

arbor vitae AR-bor VYE-tay central area of white matter in the cerebellum

arteriole ar-TEER-ee-ohl microscopic blood vessel that connects small arteries to capillaries

artery AR-ter-ee a blood vessel that carries blood away from the heart

articulation ar-tik-yoo-LAY-shun joint; point of contact between bones

atom AT-om smallest particle of an element that retains the properties of that element

atrium AY-tree-um an upper chamber of the heart; receives blood from the pulmonary or systemic system

auditory AW-di-toh-ree pertaining to the sense of hearing

auditory tube AW-di-toh-ree tube passageway that connects the nasopharynx with the middle ear

auricle AW-ri-kul curved, flexible upper portion of the ear; also, expandable flap-like structure of an atrium in the heart

axilla AK-sil-ah the armpit

axon AK-son nerve cell process that conducts impulse away from cell body

axon hillock AK-son HILL-ok funnel-shaped portion of neural cell body from which the axon extends

B

basement membrane BAYSE-ment MEM-brayne a layer of protein fibers that connects the epithelium to the underlying connective tissue

basophil BAY-so-fil white blood cell; releases histamine into damaged tissue

bicuspid valve bye-KUSS-pid valve the left atrioventricular (A-V) valve in the heart; also known as the mitral valve

bicuspids bye-KUSS-pids teeth used for crushing, mashing, and grinding; premolars

bipolar neuron bye-POH-lar NOO-ron nerve cell with two distinct processes; one dendrite, one axon

blastocyst BLASS-toh-sist early stage in embryonic development; a hollow ball of cells consisting of an inner cell mass and an outer cell mass

blastomere BLASS-toh-meer the first cleavage division that produces a pre-embryo consisting of two identical cells

bolus BOH-luss a small pastey mass of crushed and chewed food to be swallowed

brachial BRAY-kee-al pertaining to the upper limb between shoulder and elbow

brain stem brayn stem part of the brain that contains important processing centers; consists of the medulla oblongata, pons, and midbrain

bronchiole BRONG-kee-ohl small, tube-like branch of a bronchus; lacks cartilaginous supports but wall contains smooth muscle

bronchus BRONG-kuss a branch of the bronchial tree between the trachea and the bronchioles

buccal BUK-al pertaining to the cheek

bulbourethral glands BUL-boh-yoo-REE-thral glands small, mucus glands located at base of the penis; secretions lubricate the urethra

C

calcification kal-sih-fih-KAY-shun process of hardening a tissue with deposits of calcium salts

canaliculus/canaliculi kan-ah-LIK-yoo-luss/kan-ah-LIK-yoo-lye microscopic channels between cells, found in compact bone and liver; in compact bone, canaliculi allow diffusion of nutrients and wastes; in liver, bile canaliculi transport bile to bile ducts

canine tooth KAY-nyne tooth *see* cuspids

capillary KAP-i-lair-ee smallest blood vessel; connects arterioles and venules

carbohydrate kar-boh-HYE-drayt organic compound containing carbon, oxygen, and hydrogen; sugars, starches, and cellulose

carotid artery kah-ROT-id AR-ter-ee the large artery of the neck that provides a major blood supply to the brain

carpals KAR-puls wrist bones

carpus/carpal KAR-pus/KAR-pul the wrist

caudal KAW-dal the tail

CCK *see* cholecystokinin

cecum SEE-kum the pouch located at the beginning of the large intestine

cell sell the basic unit of life

cementum see-MEN-tum bone-like material covering the root of the tooth

centriole SEN-tree-ohl tiny, cylindrical organelle of a cell; involved with spindle formation during mitosis

centromere SEN-troh-meer the region where two chromatids are connected during the early stages of cell division

centrosome SEN-troh-sohm the region of cytoplasm that contains a pair of centrioles

cerebellum sair-eh-BELL-um second largest part of the brain; coordinates and refines learned movement patterns

cerebrum SAIR-eh-brum largest region of the brain; origin of conscious thoughts and all intellectual functions; controls sensory and motor integration

cervical SER-vih-kal pertaining to the neck

cervix SER-viks neck-like structure at the inferior portion of the uterus; projects into vagina

chief cells CHEEF sells cells found in the stomach; secrete pepsinogen

cholecystokinin (CCK) koh-lee-sis-toh-KYE-nin a duodenal hormone that stimulates contraction of the gallbladder and secretion of digestive enzymes from the pancreas

chondrocyte KON-droh-syte a cartilage cell

chorion KOH-ree-on a membrane consisting of the mesoderm and trophoblast; develops into a membrane of the placenta

choroid KOH-royd middle, vascular layer of the eye

chromatid KROH-mah-tid either of two daughter strands of chromosomes that are joined by a single centromere

chromatin KROH-mah-tin chromosomal material that is loosely coiled, forming a tangle of fine filaments with a grainy appearance

chromosome KROH-meh-sohm tightly compacted structures that contain coiled DNA wrapped around histone proteins; normal human body cells contain 46 chromosomes each

chyme kyme a soupy, viscous mixture of ingested substances and gastric juice first formed in the stomach

cilia SIL-ee-ah long folds of plasma membrane that contain microtubules

ciliary body SIL-ee-air-ee body a muscular structure that surrounds the perimeter of the lens of the eye and attaches to it through the suspensory ligaments

clitoris KLIT-oh-ris small female organ composed of erectile tissue located behind the junction of the labia majora

coccygeal KOKS-ih-jee-al relating to or near the coccyx

coccyx KOKS-siks the tailbone; most inferior portion of the veretebral column

cochlea KOHK-lee-ah an inner ear structure that resembles a snail shell; contains nerve endings that are essential to hearing

collagen KAHL-ah-jen the most common type of protein fiber found in connective tissues, serves to strengthen tissues

colon KOH-lon the portion of the large intestine extending from the cecum to the rectum

common bile duct KOM-mon byle dukt formed by the union of the cystic duct and the hepatic duct; carries bile from the liver and gallbladder to the duodenum

compact bone KOM-pak bone dense bone; contains osteons

compound KOM-pound a substance formed by the union of two or more elements

connective tissue koh-NEK-tiv TISH-yoo one of the four major tissue types; serve to give structural support to other tissues and organs in the body; most contain cells, protein fibers, and ground substance

cornea KOHR-nee-ah the transparent, anterior region of the sclera

corona radiata koh-ROHN-ah ray-dee-AY-tah follicular cells that surround the oocyte

coronoid KOHR-oh-noyd pertaining to certain processes of the bone; shaped like a crow's beak

corpus albicans KOHR-pus AL-bi-kans pale scar tissue in the ovaries that replaces the nonfunctional corpus luteum

corpus callosum KOHR-pus kah-LOH-sum area of the brain that links the right and left cerebral hemispheres

corpus cavernosum KOHR-pus kav-er-NO-sum two columns of erectile tissue that extend along the length of the penis

corpus luteum KOHR-pus LOO-tee-um an ovarian structure transformed from a ruptured follicle; secretes estrogen and progesterone

corpus spongiosum KOHR-pus spun-jee-OH-sum erectile body that surrounds the urethra

corpus/corpora KOHR-pus/KOHR-pohr-ah body

cortex KOHR-teks outer part of an organ

Cowper's glands (see bulbourethral glands) KOW-perz glands *see* bulbourethral glands

coxa/coxae KOKS-ah/KOKS-ee pelvic bone or hip bone

crest (as in bone marking) slightly raised narrow ridge on a bone; site for muscle attachments

crural KROOR-al refers to the anterior portion of the leg (below the knee)

cubital KYOO-bi-tal pertaining to the elbow

cuspids KUS-pids sharp, pointed teeth; canines

cystic duct SIS-tik dukt a tube that leads from the gallbladder toward the liver; unites with the common hepatic duct to form the common bile duct

cytoplasm SYE-toh-plaz-em the gel-like material between the cell membrane and the nuclear membrane

cytoskeleton sye-toh-SKEL-eh-ton an integrated network of microtubules and microfilaments in the cytoplasm that gives shape and provides support to the cell and its organelles

cytosol SYE-toh-sawl fluid portion of the cytoplasm

D

decidua basalis dih-SID-yoo-ah bah-SAY-lis area of the endometrium that develops into the maternal part of the placenta

deep away from the surface; opposite of superficial

deltoid DEL-toyd triangular shape

dendrite DEN-dryte processes of a neuron that directly respond to stimuli

dentin DEN-tin the mineralized matrix found in teeth

dermis DER-mis layer of connective tissue that lies beneath the epidermis

diaphysis dye-AF-i-sis tubular shaft of a long bone

diffusion dih-FYOO-shun the net movement of substances from an area of high concentration to an area of low concentration

disaccharide dye-SAK-ah-ryde a double unit sugar formed by bonding a pair of monosaccharides together; ex: lactose, sucrose

distal DIS-tal refers to the region or reference away from an attached base; opposite of proximal

distal convoluted tubule (DCT) DIS-tal KON-voh-loo-ted TOOB-yool the part of the nephron distal to the ascending limb of the nephron loop; site for active secretion and selective reabsorption of ions and other materials

dorsal DOR-sal pertaining to the back; posterior; opposite of ventral

ductus deferens DUK-tus DEF-er-ens a smooth muscular tube that propels sperm from the epididymis to the ejaculatory duct; also known as vas deferens

duodenum doo-oh-DEE-num first and shortest segment of the small intestine—about 10 in. long; receives chyme from the stomach and digestive secretions from the pancreas

dura mater DOO-rah MAH-ter outermost layer of the meninges

E

efferent EF-fer-ent away from; opposite of afferent

ejaculatory duct ee-JAK-yoo-lah-toh-ree dukt short passageway that allows sperm to enter the urethra

electron ee-LEK-tron negatively charged subatomic particle of an atom; located in energy shells outside the nucleus of the atom

embryo EM-bree-oh a stage of human development beginning at fertilization and ending at the start of the eighth developmental week

emulsification ee-mul-sih-fih-KAY-shun the process by which bile breaks up fats

enamel ee-NAM-el hardest manufactured substance in the body; covers the crown of the tooth

endocardium en-doh-KAR-dee-um simple squamous epithelial inner layer of the heart

endometrium en-doh-MEE-tree-um inner glandular layer of the uterine wall

endomysium en-doh-MISH-ee-um inner layer of connective tissue that surrounds each skeletal muscle fiber

endoneurium en-doh-NOO-ree-um a layer of connective tissue that surrounds individual axons

endoplasmic reticulum (ER) en-doh-PLAS-mik re-TIK-yoo-lum network of intracellular membranes that synthesizes and manufactures membrane-bound proteins

endosteum en-DOS-tee-um a membrane that lines the marrow cavity inside a bone

enzyme EN-zyme biological catalysts

eosinophil ee-oh-SIN-oh-fil a phagocytic white blood cell; numbers increase during allergic reaction

epicardium ep-i-KAR-dee-um outer covering of the heart; also called the visceral pericardium

epidermis ep-i-DER-mis outermost layer of the skin

epididymis ep-i-DID-i-miss a long, coiled and twisted tubule that lies along the posterior border of the testis; stores sperm and facilitates their maturation

epiglottis ep-i-GLOT-iss flap of elastic cartilage that folds back over the larynx during swallowing

epimysium ep-i-MISH-ee-um connective tissue layer that surrounds the entire skeletal muscle

epineurium ep-i-NOO-ree-um outermost fibrous connective tissue sheath that surrounds a peripheral nerve

epiphysis eh-PIF-i-sis expanded end of a long bone

epithelial tissue ep-i-THEE-lee-al TISH-yoo one of the four major tissue types; serves to cover exposed body surfaces and line internal cavities and passageways

erythrocyte e-RITH-roh-syte red blood cell

esophagus eh-SOF-ah-gus hollow, muscular tube that transports food and liquid from the pharynx to the stomach

estrogens ES-troh-jens steroid hormones produced by the ovaries; dominant sex hormone in females

F

facet FASS-et flat surface of a bone that forms a joint with another bone

Fallopian tubes (see uterine tubes) fal-LOH-pee-an toobs *see* uterine tubes

fascia FAY-sha a connective tissue sheath consisting of fibrous tissue and fat; unites skin to underlying tissue

fascicle FASS-i-kul a bundle found in skeletal muscle and nerves; in muscle, a bundle of skeletal muscle cells; in nerves, a bundle of axons

femoral FEM-or-al pertaining to the thigh

fetus FEE-tus the name given to the unborn young from the eighth week of pregnancy to birth

fibroblasts FYE-broh-blasts the most abundant fixed cells in connective tissue proper

fibular FIB-yoo-lar pertaining to the fibula

fimbria/fimbriae FIM-bree-ah/FIM-bree-ee finger-like projections

fissure FISH-ur deep groove

fontanel FON-tah-nel fibrous area between the cranial bones; "soft spot"

foramen/foramina foh-RAY-men/foh-RAM-i-nah a hole or passageway in bone for blood vessels and/or nerves to pass through

forearm FORE-arm the part of the arm between the elbow and the wrist

fossa FOSS-ah depression

frenulum FREN-yoo-lum small bridle; thin, flat fold of mucous membrane between two structures, such as the lingual frenulum that anchors the body of the tongue to the floor of the oral cavity.

frontal FRUN-tal toward the anterior

frontal plane FRUN-tal playne runs parallel to the long axis of the body; divides the body into anterior and posterior sections

G

G cells jee sells cells found in the stomach; secrete the hormone gastrin

gametes GAM-eets Mature male or female reproductive cells—sperm or secondary oocye.

gastric GAS-trik pertaining to the stomach

gastric juice GAS-trik joos the liquid secreted by the parietal cells and chief cells; aids in digestion

gastrin GAS-trin the hormone that stimulates the secretion of both parietal and chief cells in the stomach

genitalia jen-i-TAYL-yah the reproductive organs

glucose GLOO-kohs monosaccharide; principal source of energy in the cells

gluteal GLOO-tee-al pertaining to the buttocks

glycogen GLYE-koh-jen a polysaccharide consisting of a long chain of glucose units; found mainly in liver and skeletal muscles

Golgi complex GOL-jee KOM-pleks membranous cellular organelle that stores, alters, and distributes secretory products such as proteins; gives rise to lysosomes and secretory vesicles; also known as Golgi appartus or Golgi body

Graafian follicle GRAH-fee-en FOL-li-kul *see* tertiary follicle

gray matter gray MAT-er regions in the brain and spinal cord containing nerve cell bodies, glial cells, and unmyelinated axons

gyrus/gyri JYE-russ/JYE-rye elevated ridge

H

hard palate hard PAL-let bony roof of the mouth

haustrum/haustra HAWS-trum/HAWS-trah series of pouches in the wall of the colon; permits distention and elongation

Haversian system hah-VER-shun system
See osteon

head rounded process at the end of a long bone

hemoglobin hee-mo-GLOH-bin a protein containing iron in red blood cells; functions to transport oxygen and some carbon dioxide in the blood

hepatic duct heh-PAT-ik dukt collects bile from all of the bile ducts of the liver lobes; unite to form common bile duct

hepatic portal vein heh-PAT-ik POR-tall vane delivers blood to the liver

hepatic triads heh-PAT-ik TRYE-ads portal areas located at each of the six corners of the hepatic lobule; each triad consists of the following three structures: (1) branch of hepatic portal vein, (2) branch of hepatic artery proper, and (3) branch of the bile duct

hepatocytes heh-PAT-oh-syte liver cells

hepatopancreatic sphincter heh-PAT-oh-pan-kree-AT-ik SFINGK-ter muscular ring that surrounds the lumen of the common bile duct and the duodenal ampulla; seals off the passageway and prevents bile from entering the small intestine; also known as sphincter of Oddi

hormone HOR-mohn chemical messenger; secreted by an endocrine gland

hydroxyapatite hye-DROK-see-ap-ah-tyte the chief structural component of bone; formed by the interaction of calcium phosphate and calcium hydroxide

hypodermis hye-poh-DER-mis layer of loose connective tissue that separates the skin from underlying tissue and organs; also known as subcutaneous layer

hypothalamus hye-poh-THAL-ah-mus part of the diencephalon in the brain; contains control centers regulating autonomic functions, emotions, and hormone production

I

ileum IL-ee-um last 12-foot segment of the small intestine

ilium IL-ee-um largest coxal bone

incisors in-SYE-zors blade-shaped teeth found at the front of the mouth; used for clipping or cutting

incus IN-kuss middle ear bone; also called the anvil

inferior in-FEER-ee-or below; opposite of superior

inferior vena cava inferior VEE-nah KAY-vah one of the great veins; carries blood to the right atrium from the trunk and lower extremities

inguinal ING-gwih-nal pertaining to the groin

inner ear region of the ear that contains the receptors of equilibrium and hearing

inorganic in-or-GAN-ik pertaining to compounds that lack hydrocarbons (carbon atoms bonded to hydrogen atoms); *ex:* water, salts

intercalated discs in-TER-kah-lay-ted disks specialized regions that form connections between cardiac muscle cells

interdigitate in-ter-DIJ-ih-tayt to interlock

intermediate filaments in-ter-MEE-dee-it FIL-ah-ments part of the cell cytoskeleton; provides strength, stabilizes the positions of the organelles and transports material within the cell

interphase IN-ter-fayz longest phase in the cell life cycle when the cell prepares for division; DNA is replicated in this stage

intervertebral disc in-ter-VER-tee-bral disk pad of fibrocartilage that separates and cushions the vertebrae

iris EYE-riss part of the eye that contains blood vessels, pigment cells, and smooth muscle

ischium/ischia IS-kee-um/IS-kee-ah one of the three bones that fuse to create a coxal bone in the hip

J

jejunum jeh-JOO-num 8-foot middle segment of the small intestine between the duodenum and the ileum

K

kidney KID-nee major organ of the urinary system; filters waste products from the blood

L

labia LAY-bee-ah "lips"; skin folds on the sides of the opening to the vagina

lacrimal gland LAK-rih-mal gland almond-shaped tear gland

lacteal LAK-tee-al lymphatic capillary within a villus of the small intestine

lacuna lah-KOO-nah a small space where bone or cartilage cells are found

lamella/lamellae lah-MEL-ah/lah-MEL-ee layer of calcified bone matrix

lamina propria LAY-min-ah PRO-pree-ah the loose connective tissue component of a mucous membrane

laryngopharynx lair-ring-go-FAIR-inks portion of the pharynx lying between the hyoid bone and the entrance to the esophagus

larynx LAIR-inks cartilaginous structure that surrounds and protects the glottis and vocal cords; voice box

lateral LAT-er-al on the side; away from the midline of the body; opposite of medial

lens a part of the eye that lies posterior to the cornea; focuses the visual image on the retinal photoreceptors

leukocyte LOO-koh-syte white blood cell

ligament LIG-ah-ment a type of connective tissue that connects bone to bone

line a long, narrow strip or mark; in the skeletal system, similar to a crest

lipase LYE-payse a pancreatic enzyme that breaks down lipids

lipids LIP-idz organic molecules containing carbon, hydrogen, and oxygen, such as in fats and oils

lobule LOB-yool a small lobe

loop of Henle loop of HEN-lee *see* nephron loop

lumbar LUM-bar lower back

lumen LOO-men hollow area within a tube

lungs the pair of respiratory organs that exchanges oxygen and carbon dioxide with the blood

lymph limf fluid transported by the lymphatic system; similar to plasma, but contains lower concentration of proteins

lymph node limf node small oval organ that filters and purifies lymph before it reaches the venous system

lymphocyte LIM-foh-syte primary cell of the lymphatic system; a type of white blood cell

lysosome LYE-soh-sohm cellular organelle that contains digestive enzymes

M

macromolecules mak-roh-MOL-eh-kyools large, complex molecules made from simpler molecules

malleus MAL-ee-us one of three ear bones; also called the hammer

marrow MAIR-roh soft vascular tissue that fills the cavities of most bones

matrix MAY-triks extracellular substance of a connective tissue

meatus mee-AY-tus tubelike opening or channel

medial MEE-dee-al toward the midline longitudinal axis of the body; opposite of lateral

median plane MEE-dee-an plane a section that passes along the midline and divides the body into right and left halves

mediastinum mee-dee-as-STYE-num region of the thoracic cavity between the lungs

medulla meh-DUL-ah inner portion of an organ

medulla oblongata meh-DUL-ah ob-long-GAH-tah most inferior region of the brainstem; site of vital control centers for respiratory and cardiovascular systems

medullary cavity MED-oo-lair-ee cavity hollow space inside the shaft of the bone; marrow cavity

melanin MEL-ah-nin brown skin pigment that absorbs ultraviolet radiation

melanocytes MEL-ah-noh-sytes a fixed cell that synthesizes and stores melanin

meninx/meninges ME-ninks/meh-NIN-jeez one of the three membranes surrounding the spinal cord or brain

metabolism meh-TAB-oh-liz-em sum of all chemical activities in the body

metacarpals met-ah-KAR-puls the bones in the palm of each hand

metaphase MET-ah-fayze stage of mitosis; evident when chromatids line up at the center of the cell

microfilaments my-kroh-FIL-ah-ments slender protein strands that make up the framework of the cytoskeleton; also called actin

microtubules my-kroh-TOOB-yools hollow protein tubes found in all body cells; framework of the cytoskeleton

microvillus/microvilli my-kroh-VIL-luss/my-kroh-VIL-eye small, finger-shaped folds of the cell membrane of an epithelial cell

midbrain portion of the brainstem; provides pathways between brainstem and cerebrum

middle ear small cavity between the external ear and inner ear that houses the three small ear bones

mitochondrion/mitochondria my-toh-KON-dree-ohn/my-toh-KON-dree-ah a cellular organelle that is the site of aerobic cellular respiration; produces ATP for the cell

mitosis my-TOH-sis process of cell division that results in the formation of two daughter cells that contain the same type and number of chromosomes as the parent cell; divided into prophase, metaphase, anaphase, and telophase

molars MOHL-larz teeth with flattened crowns with prominent ridges; used for grinding and crushing

molecules MOL-eh-kyools compound consisting of two or more atoms held together by chemical bonds

monocyte MON-oh-syte a phagocytic white blood cell

monosaccharide mon-oh-SAK-ah-ryde a simple, single unit sugar, such as glucose or fructose

morula MOR-yoo-lah a solid ball of cells formed from mitotic divisions of the blastomeres

mucosa myoo-KOH-sah mucous membrane

mucous (adj.) MYOO-kuss describes lubricating secretions along the digestive, respiratory, urinary, and reproductive tracts

mucus (n.) MYOO-kuss a thick, gel-like substance secreted by glands in the digestive, respiratory, urinary, and reproductive tracts

multipolar neuron mul-tee-PO-lar NOO-ron a neuron that has several dendrites and a single axon

myelin MY-eh-lin a membranous insulation consisting of many layers of glial cell membrane around the axon of a nerve cell; improves the speed of impulse conduction

myelination my-eh-lih-NAY-shun the process of forming a myelin sheath

myocardium my-oh-KAR-dee-um the cardiac muscle in the wall of the heart

myometrium my-oh-MEE-tree-um the middle muscular layer of the uterine wall

myosin MY-oh-sin thick protein filament found in skeletal muscle cells

N

nasal NAY-zal pertaining to the nose

nasopharynx nay-zoh-FAIR-inks the superior portion of the larynx

neck a slender region, usually connected to the head of a bone

nephron NEF-ron the basic functional unit of the kidney

nephron loop NEF-ron loop a subdivision of the nephron; located between the proximal convoluted tubule and the distal convoluted tubule; also known as the loop of Henle

neurilemma noo-rih-LEM-mah the cytoplasmic covering around the axon provided by the Schwann cells; found only in the peripheral nervous system

neuroglia noo-ROG-lee-ah a class of cell types in neural tissue that provides nutrients and a supporting framework to neural tissue; does not transmit impulses

neurolemmocyte noo-ROH-lem-moh-syte a neuroglial cell that wraps itself around peripheral axons to form a protective covering

neutron NOO-tron electrically neutral subatomic particle of an atom; located within the nucleus of the atom

neutrophil NOO-troh-fill phagocytic white blood cell; produced in bone marrow

Nissl Bodies NISS-ul Bodies ribosomal clusters found in neuron cell bodies

node of Ranvier node of rahn-vee-AY gaps of exposed axon not covered by a myelin sheath; appear at regular intervals in some myelinated axons

nucleic acid noo-KLAY-ik ASS-id organic macromolecule with long, polymer structure; made up of functional units called nucleotides; examples include DNA, RNA

O

oblique oh-BLEEK diagonal

olecranon oh-LEK-rah-nohn the point of the elbow

olfaction ohl-FAK-shun the sense of smell

oocyte OH-oh-syte immature stage of the female gamete

optic chiasma OP-tik kye-AS-mah area near the diencephalon where the optic nerves cross over each other to eventually connect to opposite sides of the brain

optic nerve OP-tik nerve carries visual information from the retina of the eyes to the brain

oral OR-al pertaining to the mouth

orbital OR-bi-tal pertaining to the eye

organ OR-gan a group of tissues that perform a specific function

organelle or-gah-NELL the small internal structures in cells; divided into membranous and nonmembranous types; ex: ribosome

organic or-GAN-ik pertaining to compounds that contain hydrocarbons (carbon atoms bonded to hydrogen atoms); ex: carbohydrates, proteins

oropharynx oh-roh-FAIR-inks portion of the pharynx that extends between the soft palate and the base of the tongue at the level of the hyoid bone

osmosis os-MOH-sis movement of water across a semipermeable membrane toward the solution containing the higher solute concentration

osteoblasts OS-tee-oh-blasts cells that lay down the specialized matrix of bone; responsible for the production of new bone

osteoclasts OS-tee-oh-clasts cells that decompose the matrix of bone

osteocytes OS-tee-oh-sytes mature bone cells

osteon OS-tee-ohn the basic functional unit of mature compact bone

oval window OH-val WIN-doh membranous structure in the inner ear; separates the middle and inner ear; stapes connects to it

ovary OH-vah-ree female gonad; produces ova

ovulation ov-yoo-LAY-shun the release of an egg cell called a secondary oocyte from a tertiary follicle in the ovary

ovum/ova OH-vum/OH-vah female sex cell; egg

P

palpate PAL-payt to examine by feeling part of the body

pancreas PAN-kree-ass the slender, elongated organ that lies in the abdominopelvic cavity; has the dual function of manufacturing digestive enzymes and making hormones such as insulin

papilla pah-PILL-ah nipple-shaped mounds

parietal pah-RYE-i-tal relating to or forming the wall of an organ or cavity

parietal cells pah-RYE-i-tal sells cells located in the stomach; secrete hydrochloric acid

pectoral PEK-toh-ral pertaining to the chest; between sternal and axillary regions

pedicle PED-i-kul bony structure that attaches the vertebral arch to the body of a vertebra; means "foot"

pelvic PEL-vik relating to the pelvis

pepsin pep-SYN enzyme made by chief cells in the stomach; function to digest proteins

pepsinogen pep-SYN-oh-jen an *inactive* enzyme secreted by chief cells in the stomach; converted to the *active* enzyme pepsin during gastric digestion

perimysium pair-i-MISH-ee-um a connective tissue layer that divides the skeletal muscle into a series of compartments or bundles called fascicles

perineal pair-i-NEE-al the region between the scrotum and the anus in males and between the posterior vulva junction and the anus in females

perineurium pair-i-NOO-ree-um connective tissue sheath that surrounds a bundle of nerve fibers within a nerve and holds them together

periosteum pair-ee-OS-tee-um a fibrous layer that covers the bone

peroxisomes per-AHK-si-sohms cell organelles that absorb and neutralize toxins

phalanx/phalanges FAH-lanks/fah-LAN-jeez the finger bones and toe bones

pharynx FAIR-inks passageway that connects the nose, mouth and throat; commonly called the throat

pia mater PEE-ah MAH-ter the highly vascular, innermost layer of the meninges

placenta plah-SEN-tah the specialized organ within the uterus that supports embryonic and fetal development; also called the afterbirth

plica/plicae PLYE-kah/PLYE-kee transverse folds of the intestinal lining

polysaccharide pol-ee-SAK-ah-ryde An organic macromolucule; a complex sugar formed by bonding many monosaccharides together in a long chain, such as starch or cellulose

pons ponz part of the brainstem; connects the brainstem to the cerebellum; contains relay centers and is involved with somatic and visceral motor control

posterior pos-TEER-ee-or the back; behind; opposite of anterior

premolars pree-MOHL-larz *see* bicuspids

prepuce PREE-pus the fold of skin that surrounds the glans penis in males and clitoris in females; foreskin

progesterone pro-JES-ter-ohn female hormone produced by the ovaries; prepares the endometrium for possible implantation of the embryo and stimulates milk secretion in mammary glands.

prophase PRO-fayz initial stage of mitosis; evident when the chromosomes become visible

prostate gland PROSS-tayt small, muscular, rounded organ; produces a weak acidic secretion that contributes to about 30% of the volume of semen

proteins PRO-teens organic macromolecules with a long, complex polymer structure and a variety of functions; structural units are amino acids

proton PRO-ton positively charge subatomic particle; located within the nucleus of the atom

proximal PROK-sih-mal refers to the region or reference toward an attached base; opposite of distal

proximal convoluted tubule (PCT) PROK-sih-mal kon-voh-LOO-ted TOOB-yool part of the nephron in the kidney; located between glomerular capsule and nephron loop; primary site of reabsorption of nutrients from the filtrate

pubic PYOO-bik relating to the region of the pubis

pubis PYOO-biss the articulation between the two coxal bones

pudendum pyoo-DEN-dum region enclosing the female external genitalia; usually called the vulva

pulp cavity pulp KAV-i-tee a spongy, highly vascular cavity in the tooth

pupil PYOO-pill central opening in the iris of the eye that allows light to enter the eye

pyloric sphincter pye-LOR-ik SFINGK-ter a muscular structure that regulates the release of chyme from the stomach to the duodenum

R

ramus RAY-mus a branch, such as the vertical section of bone in the mandible that connects it to the skull

rectum REK-tum last segment of the large intestine and the end of the digestive tract

rectus REK-tus straight

renal REE-nal pertaining to the kidney

rete testis REE-tee TES-teez a maze of passageways formed from the seminiferous tubules and located within the mediastinum of the testis

retina RET-i-nah the inner layer of the eye; contains the photoreceptor cells of the eye

ribosome RYE-boh-sohm cellular organelle that is the site of protein synthesis; composed of two subunits

round window round WIN-doh a thin, membranous partition that separates the cochlear chambers from the air spaces of the middle ear

rugae ROO-gee longitudinal folds that line the inner wall of the stomach

S

sacral SAY-kral pertaining to the sacrum

sagittal SAJ-i-tal runs along with the long axis of the body

sagittal plane SAJ-i-tal plane extends from anterior to posterior and divides the body into right and left sections

sarcolemma sar-koh-LEM-ah cell membrane of a skeletal muscle cell

sarcomere SAR-koh-meer microscopic, contractile unit of a skeletal muscle cell

scapula SKAP-yoo-lah shoulder blade

Schwann cell shwon sell *See* neurolemmocyte.

sclera SKLAIR-ah the dense fibrous, white outer covering of the eye

scrotum SKRO-tum pouchlike sac, divided into two chambers; contains the testes

sebaceous gland seh-BAY-shus gland holocrine gland that secretes a waxy, oily secretion into the hair follicles

secondary oocyte SEK-on-dair-ee OH-oh-syte secondary O-o-site ovum released at ovulation

semen SEE-men fluid released from the penis during ejaculation; contains sperm and seminal fluid produced by various glands

semicircular canals sem-i-SIR-kyoo-lar kah-NALS three bony rings located in the inner ear; contain fluid-filled ducts that, in turn, contain receptors that function to maintain dynamic equilibrium

semilunar valve sem-i-LOO-nar valv pair of heart valves located at the exit point of each ventricle; each valve has three pouch-like flaps; includes the pulmonary and aortic valves

seminiferous tubules seh-mih-NIF-er-us TOOB-yools a tightly coiled structure that is located in each lobule of the testis; site where spermatozoa develops

septum/septae SEP-tum/SEP-tee a terminal partition that divides an organ

serosa seh-ROH-sah a serous membrane that covers most of the digestive tract

sinus SYE-nus a cavity or space in a tissue; a large dilated vein

soft palate soft PAL-let the portion of the roof of the mouth that lies posterior to the hard palate

soma SO-mah cell body

sperm *see* spermatozoa

spermatic cord sper-MAT-ik cord a layer of fascia, tough connective tissue and muscle surrounding the ductus deferens, and the blood vessels and nerves of the testes

spermatozoon/spermatozoa sper-mah-tah-ZOH-ohn/sper-mah-tah-ZOH-ah sperm cell; sperm cells; male gamete or sex cell

spindle fiber SPIN-dul FYE-ber a network of tubules in the cell that extend between the centriole pairs

spine slender, pointed process on a bone; site for muscle attachment

spongy bone SPUN-jee bone type of bone that has a porous network of bony plates; also known as cancellous bone

stapes STAY-peez the inner ear bone; also called the stirrup

striation STRYE-ay-shun a striped or banded appearance such as that found in skeletal muscle cells and cardiac muscle cells

submucosa sub-myoo-KOH-sah a layer of loose connective tissue in the wall of the digestive tract; large blood vessels, lymphatic vessels, and nerve fibers are found here

sulcus SUL-kus a shallow depression

superficial soo-per-FISH-al on the surface; opposite of deep

superior soo-PEER-ee-or above; opposite of inferior

suture SOO-chur the boundary between the skull bones; immovable joint

synapse SIN-aps a specialized junction where a neuron communicates with another cell (such as another neuron or a muscle cell)

synovial fluid sih-NO-vee-all FLOO-id thick and colorless lubricating fluid secreted by synovial membranes

synovial joint sih-NO-vee-all joynt freely movable joint; most numerous and anatomically most complex joint in the body

synovial membrane sih-NO-vee-all MEM-brayne connective tissue membrane that lines the spaces between bones and joints; secretes synovial fluid

T

tarsals TAR-sals ankle bones

tarsus TAR-sus ankle

telophase TEL-oh-fayz stage of mitosis; nuclear membrane forms, nuclei enlarge and chromosomes gradually uncoil and disappear

temporal TEM-poh-ral pertaining to the temples on the sides of the head, above the zygomatic arch

tendon TEN-don a cable-like, fibrous connective tissue that connects skeletal muscle to bone

tertiary follicle TER-shee-air-ee FOL-lih-kul a mature ovum in the ovary; Graafian follicle

testis/testes TES-tiss/TES-teez the male gonad that produces sperm

thalamus THAL-ah-mus located in the diencephalon of the brain; final relay point for ascending sensory information that will be sent to the cerebrum

thoracic tho-RASS-ik pertaining to the chest

thrombocyte THROM-boh-syte a blood cell that has a role in clotting: also known as a platelet

thyroid gland THY-royd gland an endocrine gland located near the trachea in the neck; produces the hormones T_3, T_4, and calcitonin

tissue TISH-yoo a group of similar cells that perform a common function

tonsil TAHN-sil a large nodule containing masses of lymphoid tissue: located in the walls of the pharynx

trachea TRAY-kee-ah long tube between the larynx and the primary bronchi that serves as air passageway; the windpipe

transverse plane TRANS-vers plane a division that lies at right angles to the long axis of the body; divides the body into superior and inferior sections

tricuspid valve try-KUS-pid valv the right atrioventricular (A-V) valve in the heart located between the right atrium and right ventricle

trochanter troh-KAN-ter large bump on a bone; site for muscle attachment

trochlea TROHK-lee-ah a bone process that resembles a pulley

tubercle TOO-ber-kal small bump on a bone;

tuberosity too-bah-ROS-i-tee bump on a bone; smaller than a trochanter

tympanic membrane tim-PAN-ik membrane the eardrum

U

umbilical cord um-BIL-i-kul cord the vascular structure that connects the fetus to the placenta

unipolar neuron YOO-nee-POH-lar NOO-ron a neuron in which the dendrite and axonal processes are continuous, and the cell body lies off to the side

ureters YOOR-eh-ters pair of long, slender, muscular tubes that transport urine from the kidneys to the urinary bladder

urethra yoo-REE-thrah a passageway that transports urine from the neck of the urinary bladder to the exterior; in males, also acts as a passageway for spermatozoa to the exterior

uterine tubes YOO-ter-in tubes the hollow, muscular tubes that transport a secondary oocyte to the uterus; also known as oviducts or Fallopian tubes

uterus YOO-ter-us hollow, muscular organ that provides mechanical protection, nutritional support and waste removal for the developing embryo

V

vagina vah-JYE-nah a muscular passageway in the female that connects the uterus with the external genitalia

vas deferens vas DEF-er-enz *see* ductus deferens

vascular vas-KYOO-ler pertaining to the blood vessels

vein vane vessels that collect blood from tissues and organs and return it to the heart

ventral VEN-trul anterior or belly side in humans; opposite of dorsal

ventricle VEN-tri-kul cavity or chamber

venule VEN-yool microscopic blood vessel that connects capillaries to small veins

villus VIL-us a fingerlike extension of mucous membrane of the small intestine

vitreous humor VIT-ree-us humor a gel-like mass that fills the posterior chamber of the eye; helps maintain shape of the eye and give support to the retina

W

white matter whyte MAT-ter nerve fibers that are covered with the myelin sheath

Y

yolk sac yohk sak the first of the extraembryonic membranes to appear; an important site for blood cell production

Z

zona pellucida ZOH-nah pel-LOO-sih-dah a region formed between the innermost follicular cells and the developing oocyte

zygote ZYE-goht a fertilized egg cell

Index